Piles and Pile Foundations

Piled foundations have traditionally been designed using empirical methods, such as the capacity based approach which the majority of codes of practice are based on. However, new research has led to substantial developments in the analysis of pile groups and piled rafts, and provided deeper insight into the mechanisms for interaction between piles, soil and rafts or caps.

Based on over 20 years of experimental and theoretical research, the book presents an overview of current piled foundation design practice under both vertical and horizontal loads as well as an examination of recent advances in the analysis and design of piled rafts. Maintaining a balance between experimental evidence, analysis and design it serves as a useful handbook for traditional pile foundations while exploring recent advances.

Professor Carlo Viggiani is Emeritus Professor of Foundation Engineering at the University of Napoli Federico II, where he had been teaching from 1975 to 2009. He has been involved as geotechnical consultant in the design and construction of earth dams, civil and industrial buildings, bridges, including the suspension bridge over the Messina Straits, tunnels and underground constructions, stabilization of landslides, and safeguarding monuments including the leaning tower of Pisa.

Professor Alessandro Mandolini graduated at the University of Napoli Federico II in 1989, and obtained his PhD in 1994 jointly from the Universities of Napoli Federico II and Roma La Sapienza. He joined the Department of Civil Engineering of the 2nd University of Napoli in 1996, and was appointed as full professor in 2010. He is member of the following ISSMGE Committees: TC212 – Deep Foundations (Core Member); ERTC3 – Piles (Core Member); ERTC10 – Evaluation of Eurocode 7.

Professor Gianpiero Russo graduated at the University of Napoli Federico II in 1992 and obtained his PhD in Geotechnical Engineering jointly from the Universities of Napoli Federico II and Roma La Sapienza in 1996. He joined the Geotechnical Engineering Department of the University of Napoli Federico II in 2000 and is actually teaching Foundation Engineering. Currently he is member of the ISSMGE Committee TC204 – Underground Construction.

Piles and Pile Foundations

Carlo Viggiani, Alessandro Mandolini and
Gianpiero Russo

CRC Press
Taylor & Francis Group
Boca Raton London New York

CRC Press is an imprint of the
Taylor & Francis Group, an **informa** business

A SPON PRESS BOOK

CRC Press
Taylor & Francis Group
6000 Broken Sound Parkway NW, Suite 300
Boca Raton, FL 33487-2742

First issued in paperback 2019

© 2012 by Taylor & Francis Group, LLC
CRC Press is an imprint of Taylor & Francis Group, an Informa business

No claim to original U.S. Government works

ISBN-13: 978-0-415-49066-5 (hbk)
ISBN-13: 978-0-367-86544-3 (pbk)

British Library Cataloguing in Publication Data
A catalogue record for this book is available from the British Library

Library of Congress Cataloging in Publication Data
Viggiani, Carlo.
  Piles and pile foundations/Carlo Viggiani, Alessandro Mandolini, Gianpiero
  Russo. – 1st ed.
  p. cm.
  1. Piling (Civil engineering) I. Mandolini, Alessandro. II. Russo, Gianpiero. III.
  Title.
  TA780.V54 2011
  624.1'54–dc22

                                                                  2011007157

ISBN: 978-0-415-49066-5 (hbk)
ISBN: 978-0-203-88087-6 (ebk)

Typeset in Sabon
by Wearset Ltd, Boldon, Tyne and Wear

**Visit the Taylor & Francis Web site at
http://www.taylorandfrancis.com**

**and the CRC Press Web site at
http://www.crcpress.com**

To our wives in appreciation of their patience and support

Contents

Illustrations

Figures

Tables

Introduction

Piles have been used by humankind for foundation purposes since prehistoric times, but in the last few decades the development in equipment and installation techniques, and the pressure towards constructing in areas with poor subsoil properties, have led to spectacular progress in the piling industry.

According to Van Impe (2003) bored and CFA piles account for 50% of the world pile market, while the remainder is mainly covered by driven (42%) and displacement screw piles (6%). Different proportions may be found locally; for instance displacement screw piles represent about 60% of the total installed yearly in Belgium while bored and CFA piles reach more than 90% in Italy.

Since the behaviour of piles depends markedly on the effects of the installation technique, their design is a complex matter which, although based on the concepts of soil mechanics, inevitably relies heavily on empiricism.

The regional design practice in different countries develops along different paths under the push of the local market. In an international seminar on the design of axially loaded piles (De Cock and Legrand 1997), organized with the aim of reviewing practice in European countries, it has been confirmed that the common approach to the design of piles is essentially based on semi-empirical rules, sometimes calibrated against purposely performed load tests. It appears that there is room for improvement in the present practice.

The aim of this book is twofold. First, it reports an overview of the present design practice for pile foundations, trying to summarize some of the findings of recent research and to evaluate the above practice in the light of this research.

Second, it is shown that the deeper insight of piled foundations behaviour achieved in the last decades may be used for a more rational and economic design of piled rafts.

Any engineering problem can be tackled following one or the other of two rather different approaches. They can be illustrated using, as an analogy, two different ways of producing pieces of furniture.

One is the modern industrial way, exemplified by the products of organizations like IKEA. The industrial process allows recourse to excellent designers, advanced techniques and good materials; the result is a variety of objects, widely available on the market at relatively cheap prices. They are rather successful especially among the younger generations. You can get valuable items, provided you are able to choose the right one among the many suggestions.

The second way is that of the expert cabinet maker who restores ancient pieces of furniture in his workshop and reproduces them. He uses materials such as alcohol,

French polish, isinglass and rosewood veneer. The simplicity of tools is compensated for by experience and ability; he can produce authentic masterpieces, but in a limited number and hence not readily available.

IKEA may be assimilated to a modern and efficient engineering organization, and its industrial processes to the finite element codes that make our professional life simpler and allow us to produce reports with impressive multicoloured diagrams. On the other side, the expert craftsman is similar to the old-fashioned engineer, who still uses pencil and graph paper, superintends personally site and laboratory investigations and seldom ventures using a spreadsheet, preferring his trustworthy pocket calculator.

In principle, either of the two methods may be used successfully.

A proper use of the simple craftsman's methods requires a deep insight into the mechanics of the phenomena and a good deal of experience, to make an intelligent use of simple models and empiricism. This appears to be a significant shortcoming, since experience is not readily available to everyone.

On the other side, the sophisticated modern computer codes appear (and are widely believed to be) rather objective, as if the role of insight and experience was vanishing. It may be so, but anyone trying to apply FEM to practical design problems soon discovers how sensitive are the results to apparently minor modelling or computational details or to small variations of some esoteric parameter in a constitutive equation. "A difficulty of advanced numerical analysis" claim Vaughan *et al.* (2004) "is that the knowledge and skill required to perform numerical analyses is substantially greater than for the simple methods of approximate analysis to which we have become accustomed. This is at a time when there is a shortage of trained engineers in general and of geotechnical engineers in particular."

The present day practice of foundation design is suited for simple models and tools such as design charts, well-tempered empiricism and so on. These tools are summarized in Parts II and III of this book.

Sound engineering judgement, which is needed for a successful use of these simple methods, is not necessarily based only on personal experience. First of all, it must be firmly grounded in mechanics. Terzaghi (1943) claims that "theory is the language by means of which lessons of experience can be clearly expressed"; Peck that "theory and calculations are not substitute for judgement, but are the basis for sounder judgement" (Dunnicliff and Young 2006). And Immanuel Kant (1793) authoritatively states: "There is nothing more practical than a good theory." Accordingly, a simple mechanistic framework of classical soil mechanics is reported at the beginning of this book, and recalled throughout as a support to the judgement. Furthermore, each equation presented is accompanied by its derivation, or at least an indication of how it is obtained.

Second, quite a lot of experience has been collected on piled foundations, and made available through papers, reports and books. A critical scrutiny of such evidence is another important factor in the formation of sound engineering judgement. For each argument, a discussion of the available experimental evidence and of the main conclusions that can be drawn from it is thus attempted.

Third, following the practice of the best restaurants, the menu presented is ample and varied but for each item there are "suggestions of the chef".

A common design assumption for pile foundations is that of neglecting the contribution of the cap, and assuming that the whole load is transmitted to the soil through the piles. Most codes and regulations suggest or prescribe this approach. The assumption may be reasonable for small pile groups, but in the case of medium and large piled rafts it is unduly conservative. In fact, a proper consideration of the contribution of the raft allows significant economy.

Furthermore, present design is capacity based, in the sense that a prescribed coefficient of safety against a bearing capacity failure has to be warranted. Again, most codes and regulations suggest or prescribe this approach. Alternative design requirements could be based on a limiting value of the absolute or differential settlement. Adopting such an alternative approach, an improvement in design practice and substantial economy can be achieved for medium and large piled rafts.

Part IV of this book is devoted to these subjects. Again, emphasis is on the development of an insight of the mechanical behaviour and the factors controlling it, and simple models and methods are presented as far as possible. Some concessions to more complex and sophisticated numerical analyses, such as Boundary Element or Finite Element, is, however, unavoidable even if kept to an absolute minimum.

Simple models imply the renunciation to a detailed description, complex numerical analyses to a synthetic understanding. Since both renunciations are too grave to be tolerated, it is essential that the two strategies be considered as complementary rather than alternative.

Part I
General

1 Drained and undrained conditions; total and effective stress analysis

1.1 Stress and pore pressure

In geotechnical engineering the soil is treated as a continuum, in order to take advantage of the tools of continuum mechanics. A geometrical point in the equivalent continuum corresponds to an elementary volume of soil, large enough to include a significant number of particles and voids but still small enough in comparison to the scale of engineering problems. Irrespective of the ideal geometrical point being located inside a particle or in a void, it is thus possible to define in each point a stress tensor whose components are:

$$
\begin{matrix}
\sigma_x & \tau_{xy} & \tau_{xz} \\
\tau_{yx} & \sigma_y & \tau_{yz} \\
\tau_{zx} & \tau_{zy} & \sigma_z
\end{matrix}
\tag{1.1}
$$

The stress defined by Eq. 1.1 acts on both pore fluid (if any) and solid skeleton and is called total stress. For the sake of simplicity, we refer to either a dry soil or a fully saturated one; this is not a significant limitation for the large majority of foundation problems, since fine-grained natural soils are practically always saturated, at least in temperate regions. Coarse-grained soils are saturated if they are located below the groundwater table; if above, from the viewpoint of their mechanical behaviour they can be considered as dry.

Assuming a frame of reference with the axis z vertical and directed upwards, in any point of a water-saturated porous medium we can define also (Figure 1.1):

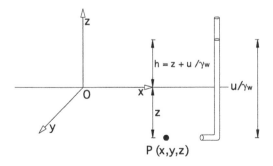

Figure 1.1 Pore water pressure and piezometric head in a saturated soil mass.

- the pressure of the water filling the voids, or pore pressure u; obviously pore pressure is nil for dry soils;
- the geometrical elevation z;
- the pressure head u/γ_w, where γ_w is the unit weight of water;
- the total head or piezometric level $h = (z + u/\gamma_w)$.

The kinetic energy term $v^2/2g$, where g is the acceleration of gravity, may be neglected, being the velocity of water v extremely small in seepage phenomena.

1.2 Permeability and seepage

According to d'Arcy (1856) the volume q of water flowing perpendicular to an area A in unit time may be expressed as:

$$q = -kA \frac{\partial h}{\partial s}$$

where s is the direction of flow, normal to the area A. The coefficient of permeability k has the dimensions of a velocity $[LT^{-1}]$ and is a characteristic of the porous medium and the seeping fluid. The negative sign means that the flow is positive in the sense of decreasing h.

Dividing both members by A we get the so-called d'Arcy's law (d'Arcy 1856):

$$V = -k \frac{\partial h}{\partial s}$$

The seepage velocity V is not the actual velocity of the water within the pores of the soil, since it is obtained by dividing the flow quantity by the whole area, while obviously water cannot flow through that part of A that is taken up with soil solids.

In a 3D flow field and in a medium with isotropic permeability the law may be expressed in vectorial form as follows:

$$\vec{V} = k \cdot grad(-h)$$

or in the equivalent scalar form:

$$V_x = -k \frac{\partial h}{\partial x}; \qquad V_y = -k \frac{\partial h}{\partial y}; \qquad V_z = -k \frac{\partial h}{\partial z}$$

If we assume that no volumetric deformation of the solid skeleton occurs (or, in other words, that the pore volume does not change), the quantity of water entering any elementary volume of the soil must be equal to that leaving it. This may be expressed imposing that the divergence of the seepage velocity is nil:

$$div \vec{V} = \frac{\partial V_x}{\partial x} + \frac{\partial V_y}{\partial y} + \frac{\partial V_z}{\partial z} = 0$$

and hence:

$$-k\left(\frac{\partial^2 h}{\partial x^2} + \frac{\partial^2 h}{\partial y^2} + \frac{\partial^2 h}{\partial z^2}\right) = -k\Delta^2 h = -\frac{k}{\gamma_w}\Delta^2 u = 0 \tag{1.2}$$

The value of the coefficient of permeability of natural soils depends essentially on their grain size and ranges in a very broad interval: from 10^{-11} m/sec to 10^{-1} m/sec passing from homogeneous clay to sand and gravel. This extreme variation is responsible for the significant differences in the mechanical behaviour of fine-grained (clay, silt) and coarse-grained (sand, gravel) materials. A seepage process requiring a short time (say some minutes to some hours) to be completed in a coarse sand layer may require a time as long as tens or even hundreds of years in a large clay mass.

When a saturated soil is subjected to a system of loads, the volume of its pores tends to increase or to decrease; in terms of continuum mechanics, there is the tendency to a volumetric deformation. The water and the solid particles being very nearly incompressible in comparison to the soil skeleton, a volumetric deformation is possible only by expelling or adsorbing water with a transient flow.

In a coarse-grained material this flow requires a very short time; accordingly, except for the practically instantaneous transient stage, the characteristics of the water motion at a point (fluid pressure u, seepage velocity V) do not change in time and depend only on the hydraulic boundary conditions. If the boundary conditions are variable in the time (e.g. cycles of filling and emptying of a reservoir), the resulting transient motion may be treated as a succession of steady states. In terms of fluid mechanics, this is called a steady state or permanent motion; since there are no stress variations, there are no volume changes and hence the hypothesis of incompressible solid skeleton may be accepted. In terms of soil mechanics, this is called a *drained condition*.

In a drained condition the pore pressure and the stress acting on the solid skeleton are uncoupled. The field of the piezometric head h (and pore pressure u) may be obtained by solving Eq. 1.2 with the proper boundary conditions. To this end, analogical, numerical or analytical techniques are available. The subject is too broad to be even touched upon here; the interested reader may refer to the classical treatises by Muskat (1953), Polubarinova-Kochina (1956), Aravin and Numerov (1964), Cedergreen (1967) and Bear (1972).

1.3 Principle of effective stress

At the beginning of the twentieth century the theoretical framework of mechanics was available to engineering, with its corollary of methods of analysis and solutions. Solid mechanics, and especially the theory of elasticity, was the basis for modern structural engineering; fluid mechanics for hydraulic engineering.

A number of contributions relevant to geotechnical problems were already available (Coulomb 1776; d'Arcy 1856; Boussinesq 1885), but soil mechanics and geotechnical engineering did not yet flourish. The clue for understanding the fundaments of soil mechanics was the concept of effective stress and the statement of the so-called principle of effective stress.

This principle, commonly attributed to Terzaghi (1924), consists of two parts:

1   a definition of effective stress σ' as the difference between the total normal stress σ acting on the complex (water + soil particles), and the pressure u of the pore water:

$$\sigma' = \sigma - u$$

2   the statement that the behaviour of the soil (deformation, failure) is controlled by the effective stress.

The principle of effective stress is generally introduced by the well-known spring and piston model of one-dimensional consolidation. The spring and piston model is simple and clear; it is to underline, however, that one-dimensional consolidation is but a special case that cannot be generalized. In the majority of foundation problems stress, deformation and flow are three-dimensional and, if this circumstance is not considered, some phenomena cannot be explained (just to quote an example, the immediate or undrained settlement of a foundation). The effective stress tensor is obtained by subtracting from the total stress tensor the pore pressure multiplied by the unit tensor. In the effective stress tensor the normal stress components are thus obtained by subtracting the pore pressure from the corresponding total normal stress, while the shear stress is unchanged since water cannot sustain shear. The components of the effective stress tensor are:

$$
\begin{matrix}
\sigma_x - u & \tau_{xy} & \tau_{xz} \\
\tau_{yx} & \sigma_y - u & \tau_{yz} \\
\tau_{zx} & \tau_{zy} & \sigma_z - u
\end{matrix}
$$

In engineering problems (failure problems, such as the evaluation of the bearing capacity of a foundation; deformation problems, such as the evaluation of the settlement of a foundation) the total stress may be generally obtained by equilibrium and compatibility conditions; to evaluate the effective stress, it is thus necessary to know the pore pressure.

In any point of a soil, the indefinite equations of equilibrium have to be satisfied. For static problems (i.e. in the absence of inertia forces) and in terms of total stress they are written:

$$\frac{\partial \sigma_x}{\partial x} + \frac{\partial \tau_{xy}}{\partial y} + \frac{\partial \tau_{xz}}{\partial z} = 0$$

$$\frac{\partial \tau_{yx}}{\partial x} + \frac{\partial \sigma_y}{\partial y} + \frac{\partial \tau_{xz}}{\partial z} = 0 \qquad (1.3)$$

$$\frac{\partial \tau_{zx}}{\partial x} + \frac{\partial \tau_{zy}}{\partial y} + \frac{\partial \sigma_z}{\partial z} + \gamma_{sat} = 0$$

where γ_{sat}, the unit weight of the saturated soil, is the only non-zero volume force acting in the z direction. Eq. 1.3 describes the equilibrium of the solid skeleton and the water.

Remembering that $\sigma_i = \sigma'_i + u$ and $u = (h - z)\gamma_w$, Eq. 1.3 may be written:

$$\frac{\partial \sigma'_x}{\partial x} + \frac{\partial \tau_{xy}}{\partial y} + \frac{\partial \tau_{xz}}{\partial z} + \gamma_w \frac{\partial h}{\partial x} = 0$$

$$\frac{\partial \tau_{yx}}{\partial x} + \frac{\partial \sigma'_y}{\partial y} + \frac{\partial \tau_{xz}}{\partial z} + \gamma_w \frac{\partial h}{\partial y} = 0 \qquad (1.4)$$

$$\frac{\partial \tau_{zx}}{\partial x} + \frac{\partial \tau_{zy}}{\partial y} + \frac{\partial \sigma'_z}{\partial z} + \gamma_w \frac{\partial h}{\partial z} + \gamma' = 0$$

where $\gamma' = \gamma_{sat} - \gamma_w$ is the submerged unit weight of the soil. Eq. 1.4 describes the equilibrium of the solid skeleton of the soil, acted upon by the effective stress and by a system of volume forces including the buoyant unit weight γ' and the forces $\gamma_w \dfrac{\partial h}{\partial i}$, the viscous drag exerted on the soil skeleton by the seeping fluid, called seepage forces.

In drained conditions, stress and pore pressure are uncoupled; the field of h is obtained separately by solving Eq. 1.2; the seepage forces in Eq. 1.4 are to be considered as known terms.

If the ground water regime is hydrostatic (the simplest case of steady state motion) the piezometric head $h =$ cost; the seepage forces vanish and again the only non-zero volume force is the submerged unit weight γ'.

1.4 Consolidation

As noted above, in a fine-grained material the transient flow may require a time as long as tens or even hundreds of years; for this reason it is necessary to analyse it, removing the hypothesis that the solid skeleton is incompressible. It follows that there is a coupling between stress and deformation in the solid skeleton on one side and pressure and seepage velocity in the water on the other. The process of transient flow of water and contemporary development of volumetric deformation is called consolidation.

If we assume that both the solid particles and the fluid are incompressible, the condition of continuity of the fluid may be expressed by imposing that the quantity of fluid entering or leaving a volume of soil in a time interval be equal to the volumetric deformation of the soil, i.e. to the variation of the pore volume. The quantity of fluid may be expressed as $div\vec{V} \cdot dxdydzdt$; the volumetric deformation by the first invariant of the strain tensor $(\varepsilon_x + \varepsilon_y + \varepsilon_z)dxdydz$; the continuity requires that:

$$div\vec{V} = \frac{\partial}{\partial t}\left(\varepsilon_x + \varepsilon_y + \varepsilon_z\right)$$

Assuming that the solid skeleton may be treated as an elastic solid with Young modulus E and Poisson's ratio v, it follows that:

$$\frac{\partial}{\partial t}\left(\varepsilon_x + \varepsilon_y + \varepsilon_z\right) = \frac{1-2v}{E}\frac{\partial}{\partial t}\left(\sigma'_x + \sigma'_y + \sigma'_z\right) = \frac{1-2v}{E}\frac{\partial}{\partial t}\left(\sigma_x + \sigma_y + \sigma_z - 3u\right)$$

Calling $T = \sigma_x + \sigma_y + \sigma_z$ the first invariant of the total stress tensor, and remembering the expression of the divergence, we obtain finally:

$$\frac{kE}{3\gamma_w\left(1-2v\right)}\Delta^2 u = \frac{\partial u}{\partial t} - \frac{1}{3}\frac{\partial T}{\partial t} \qquad\qquad (1.5)$$

Eq. 1.5 is the equation of the consolidation of a poroelastic body saturated by an incompressible fluid, and was obtained independently by Biot (1941) and Mandel (1950). The coupling of pore pressure and stress requires that it be solved, satisfying contemporarily the equilibrium and compatibility conditions.

1.5 Some examples

1.5.1 In situ *stress*

Let us consider a soil limited by a horizontal ground surface. If the lateral extent of the soil mass is large in comparison to the depth of interest, any vertical axis may be regarded as an axis of symmetry, and hence no shear stress can act on vertical planes. This means that the principal stress directions are vertical and horizontal. The total normal stress σ_z acting on a horizontal plane at the depth z is given by the weight of the overburden material. The relevant expressions are reported in Figure 1.2 for different positions of the ground water table.

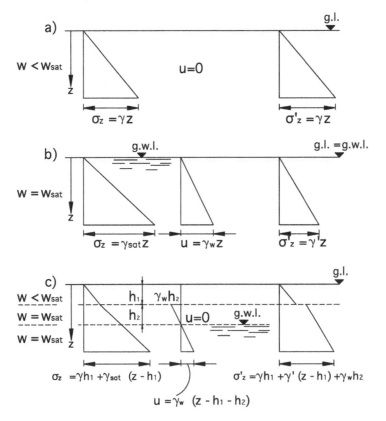

Figure 1.2 In situ total and effective stress and pore water pressure: (a) dry soil; (b) saturated soil with ground water table at the soil surface; (c) ground water table at depth and capillary fringe.

With reference to Figure 1.2c it is to be remembered that in fine-grained soil (clay, silt) the capillary rise may reach some tens of metres; for this reason, at least in temperate regions, the fine-grained soils are practically always saturated. In coarse-grained soils (sand, gravel), on the contrary, the capillary rise amounts to a few centimetres and therefore the capillary zone is generally neglected.

The horizontal effective stress may be expressed as a fraction of the corresponding vertical effective stress:

$$\sigma'_x = \sigma'_y = k_o \sigma'_z$$

If the lateral extent of the soil mass is large in comparison to the depth of interest, again any vertical axis may be regarded as an axis of symmetry, and hence the horizontal strain must be equal to zero. The ratio k_o between the horizontal and vertical effective principal stress in condition of zero horizontal strain is called coefficient of earth pressure at rest.

If the solid skeleton is assumed to be an elastic material with Young modulus E and Poisson's ratio v, we can write:

$$\varepsilon_x = \frac{1}{E}\left[\sigma'_x - v\left(\sigma'_y + \sigma'_z\right)\right] = 0$$

Taking into account that $\sigma'_x = \sigma'_y$, one obtains:

$$k_o = \frac{\sigma'_x}{\sigma'_z} = \frac{v}{1-v}$$

Real soils are not linearly elastic, and the value of k_o is markedly affected by the previous stress history of the soil. In normally consolidated deposits, the expression suggested by Jaky (1936):

$$k_o = 1 - \sin\varphi'$$

is supported by a broad experimental evidence. Combining the two expressions, one obtains:

$$v = \frac{1-\sin\varphi'}{2-\sin\varphi'}$$

In overconsolidated soils, Kulhawy *et al.* (1989) suggest to evaluate k_o with the expression:

$$k_{o(OC)} = \left(1 - \sin\varphi'\right)OCR^{\sin\varphi'}$$

where $k_{o(OC)}$ is the value of k_o for the overconsolidated soil and OCR is the overconsolidation ratio, i.e. the ratio between the maximum past effective vertical stress and the present *in situ* vertical effective stress.

1.5.2 *Piping*

In a soil interested by a steady state vertical seepage, directed upwards and merging in a free surface (Figure 1.3), the seepage force is:

$$\gamma_w \frac{\partial h}{\partial z} = \gamma_w i$$

where $i = H/L$ is the hydraulic gradient. If the seepage force equals the submerged unit weight γ' of the soil, effective stress vanishes and the soil is freely removed by the seeping water. This phenomenon is known as *piping*. For piping to occur, the upward gradient must reach a critical value i_c that may be evaluated by imposing that $i_{c \cdot \gamma w} = \gamma'$. One obtains:

$$i_c = \frac{\gamma'}{\gamma_w}$$

A practical situation in which piping may occur is represented in Figure 1.4.

1.5.3 *One-dimensional consolidation*

If the consolidation (deformation, water flow) occurs only in one direction, say along z axis, we have:

$$\varepsilon_x = \varepsilon_y = 0; \quad \frac{\partial h}{\partial x} = \frac{\partial h}{\partial y} = 0; \quad \sigma'_x = \sigma'_y = k_o \sigma'_z = \frac{v}{1-v} \sigma'_z;$$

$$T = \sigma'_z \left(1 + 2\frac{v}{1-v}\right) + 3u$$

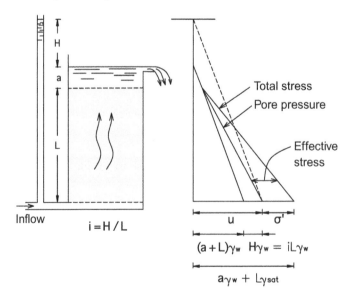

Total stress

Pore pressure

Effective stress

Inflow

$i = H/L$

u σ'

$(a + L)\gamma_w$ $H\gamma_w = iL\gamma_w$

$a\gamma_w + L\gamma_{sat}$

Figure 1.3 Upward seepage and piping.

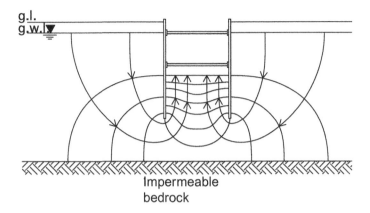

g.l.
g.w.l

Impermeable
bedrock

Figure 1.4 Possible piping at the bottom of an excavation.

Under these conditions, and if the external load does not change with the time, $\frac{\partial \sigma_z'}{\partial t} = 0$ and Eq. 1.5 reduces to:

$$\frac{kE(1-v)}{\gamma_w(1+v)(1-2v)}\frac{\partial^2 u}{\partial z^2} = \frac{\partial u}{\partial t} \tag{1.6}$$

The constrained modulus or oedometric modulus $E_{oed} = \frac{\sigma_z'}{\varepsilon_z}$ with $\varepsilon_x = \varepsilon_y = 0$, for a poroelastic medium may be shown to be expressed by:

$$E_{oed} = E\frac{1-v}{(1+v)(1-2v)} \tag{1.7}$$

Eqs 1.6 and 1.7 give:

$$\frac{kE_{oed}}{\gamma_w}\frac{\partial^2 u}{\partial z^2} = \frac{\partial u}{\partial t}$$

that is the well-known Terzaghi's 1D consolidation equation, with the coefficient of consolidation $c_v = \frac{kE_{oed}}{\gamma_w}$.

1.6 Undrained conditions: analysis in terms of total stress

To determine the evolution in time of the state of stress and deformation in the soil a complete solution of the problem of consolidation would be required. Fortunately, this is not always necessary. From the viewpoint of foundation engineering, in the majority of problems only the analysis of two particular situations is required: the initial and the final one.

Let us imagine that the consolidation process starts at $t = 0$, for instance, with the application of the load to a foundation. At the starting time, the permeability being finite (and for clayey soils actually very small), no flow of water may occur; this ideal initial condition is called *undrained condition*.

In engineering problems, the loads are actually applied in a finite time, ranging from days to months, or even years, and therefore during load application some consolidation does occur. The consolidation time (i.e. the time needed for a substantial completion of the consolidation process) of a clay layer of significant thickness, however, is often much longer (decades or centuries) than the duration of the construction. The simplifying assumptions of perfectly undrained initial condition may thus be accepted, considering also that for foundation problems it is a conservative one.

In a saturated porous medium, with incompressible soil particles and water, no volume deformation may occur in undrained conditions and only shear deformations without volume changes are possible. This requires that:

$$\left(\varepsilon_x + \varepsilon_y + \varepsilon_z \right)_{t=0} = 0$$

In an elastic material, the unit volume variation depends only on a corresponding variation of the mean effective stress; we may write thus:

$$\left(\sigma'_x + \sigma'_y + \sigma'_z \right)_{t=0} = \left(\sigma_x + \sigma_y + \sigma_z - 3u \right)_{t=0} = 0$$

where total and effective stress are intended as increments and pore pressure as excess pore pressure. The undrained excess pore pressure u_o induced by an increment of total stress may be obtained as:

$$u_o = \frac{\sigma_x + \sigma_y + \sigma_z}{3} \tag{1.8}$$

Eq. 1.8 is valid for an elastic and isotropic material, in which the volumetric deformation depends only on the variations of normal stress, and shear stress and volumetric strains are uncoupled. In real soils, on the contrary, such a coupling occurs, and therefore Eq. 1.8 has to be modified to account for it. The most widespread expression of the excess pore pressure in undrained conditions is due to Skempton (1954):

$$u_0 = B \left[\Delta\sigma_3 + A \left(\Delta\sigma_1 - \Delta\sigma_3 \right) \right]$$

where $\Delta\sigma_1$, $\Delta\sigma_3$ are the increment of major and minor principal total stress and A, B are called pore pressure coefficients; for saturated soils $B = 1$.

In principle, we have thus the tools to analyse engineering problems (deformation, failure) in undrained conditions and in terms of effective stress. In practice, however, the analysis is cumbersome and uncertain.

It has been observed that, in the undrained conditions, the water saturated porous material behaves, in terms of total stress, as a continuous body characterized by mechanical properties that have to be determined by undrained tests. It is possible to define an undrained modulus E_u, an undrained Poisson ratio $v_u = 0.5$ (incompressible material), an undrained strength c_u with $\varphi_u = 0$. In other words, any deformation or failure problem under undrained conditions may be analysed very simply in terms of total stress referring to an incompressible, purely cohesive equivalent material. If the soil properties are correctly determined, the analysis of an undrained problem in terms of total stress or effective stress should give the same results.

1.7 Equivalence of the analysis in terms of total and effective stress in undrained conditions

Two simple examples of the equivalence of total and effective stress analyses under undrained conditions are reported in the following.

As a problem of deformation, let us consider the undrained or immediate settlement of a rigid circular plate of diameter D subjected to a load system with an axial resultant equal to Q and resting at the surface of a poroelastic saturated half space (Figure 1.5). Along the axis of the foundation the stress increments induced by the load Q are given by the expressions (Fischer 1957):

$$\sigma_z = \frac{1}{2} q \sin^2 \alpha \left(1 + 2\cos^2 \alpha\right)$$

$$\sigma_x = \sigma_y = \frac{1}{4} q \sin^2 \alpha \left(1 + 2v - 2\cos^2 \alpha\right)$$

(1.9)

where $q = 4Q/\pi D^2$ is the average unit stress acting on the foundation and $\alpha = tg^{-1} \dfrac{D}{2z}$

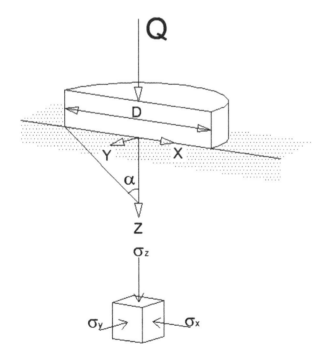

Figure 1.5 Rigid circular plate resting on an elastic half space and subjected to a vertical load Q.

Under undrained conditions, the increments of total stress σ_{zo}, σ_{xo} may be obtained by putting $v = 0.5$ in Eq. 1.9:

$$\sigma_{zo} = \frac{1}{2} q \sin^2 \alpha \left(1 + 2\cos^2 \alpha\right)$$

$$\sigma_{xo} = \sigma_{yo} = \frac{1}{2} q \sin^4 \alpha$$

(1.10)

Furthermore, in a poroelastic medium where volumetric and distortional strain are connected respectively to normal and shear stress only, the undrained excess pore pressure u_o is given by Eq. 1.8 and hence, from Eq. 1.10:

$$u_o = \frac{1}{2} q \sin^2 \alpha$$

(1.11)

From Eqs 1.10 and 1.11 we obtain:

$$\sigma'_{zo} = \sigma_{zo} - u_o = q \sin^2 \alpha \cos^2 \alpha$$

$$\sigma'_{xo} = \sigma'_{yo} = \sigma_{xo} - u_o = -\frac{1}{2} q \sin^2 \alpha \cos^2 \alpha$$

(1.12)

Finally, for an elastic solid skeleton with Young modulus E and Poisson ratio v in terms of effective stress, the elastic constants of the saturated medium under undrained conditions and in terms of total stress are $v_u = 0.5$ (incompressible medium) and $E_u = \dfrac{3}{2(1+v)} E$.

Now we have all the elements to compute the undrained settlement w_o of the footing both in terms of total and effective stress. In terms of effective stress we have:

$$w_o = \int_0^\infty \varepsilon_{zo} dz = \frac{1}{E} \int_0^\infty \left(\sigma'_{zo} - v\langle \sigma'_{xo} + \sigma'_{yo}\rangle\right) dz = \frac{q}{E}(1+v)\int_0^\infty \sin^2 \alpha \cos^2 \alpha \cdot dz$$

Being $dz = -\dfrac{D}{2\sin^2 \alpha} d\alpha$, one obtains:

$$w_o = \frac{qD}{2E}(1+v)\int_0^{\pi/2} \cos^2 \alpha d\alpha = (1+v)\frac{\pi q D}{8E}$$

In terms of total stress we have:

$$w_o = \frac{1}{E_o} \int_0^\infty \left[\sigma_{zo} - 0.5\left(\sigma_{xo} + \sigma_{yo}\right)\right] dz = \frac{q}{2E_o}\int_0^\infty \left[\sin^2 \alpha\left(1 + 2\cos^2 \alpha - \sin^2 \alpha\right)\right] dz$$

$$= \frac{qD}{4E_o} \int_{\pi/2}^0 3\cos^2 \alpha d\alpha = \frac{3}{16}\frac{\pi q D}{E_o} = (1+v)\frac{\pi q D}{8E}$$

The immediate or undrained settlement may be thus evaluated both in terms of total or effective stress, provided the proper values of elastic constants and pore pressure are adopted.

The equivalence between analyses in terms of total and effective stress in undrained conditions for a failure problem will be demonstrated by the simplified analysis of a vertical cut (see Figure 1.6a and Bishop and Bjerrum 1961). To simplify the mathematics, it is assumed that the undrained strength c_u does not vary with depth and that the effective cohesion is negligible ($c'=0$). A plane failure surface without tension cracks will be considered. The critical height H under these conditions is known to be $H=4c_u/\gamma_{sat}$, where γ_{sat} is the saturated unit weight of the soil.

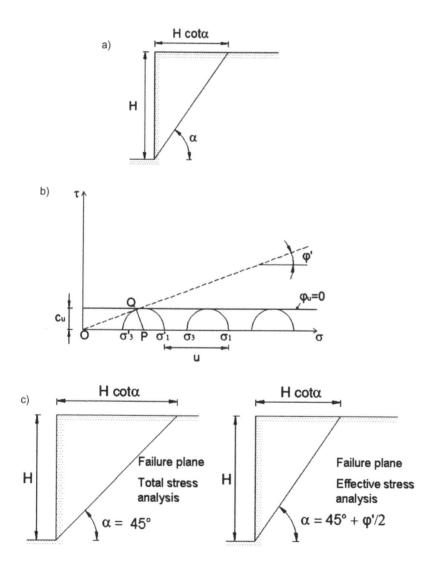

Figure 1.6 Simplified analysis of a vertical cut (after Bishop and Bjerrum 1961).

In terms of total stress the factor of safety is obtained by the expression:

$$FS = \frac{c_u l}{W \sin \alpha} = \frac{2c_u}{\gamma_{sat} H \sin \alpha \cos a}$$

being $W = \frac{\gamma_{sat} H^2}{2 tg\alpha}$ the weight of the soil above the slip plane and $l = H/\sin \alpha$ the length of the slip surface.

For $\alpha = 45°$, $\dfrac{dFS}{d\alpha} = 0$; the lowest value of FS is thus:

$$FS = \frac{4c_u}{\gamma_{sat}}$$

Substituting $4c_u/\gamma_{sat}$ for H we obtain $FS = 1$.

In terms of effective stress, from the Mohr diagram of Figure 1.6b we obtain:

$$\frac{\sigma_1 - \sigma_3}{2} = \left[(\sigma_1 - u) - \frac{\sigma_1 - \sigma_3}{2} \right] \sin \varphi'$$

Putting $\dfrac{\sigma_1 - \sigma_3}{2} = c_u$ and rearranging, we obtain:

$$u = \sigma_1 - c_u \frac{1 + \sin \varphi'}{\sin \varphi'} \tag{1.13}$$

For a plane slip surface, the expression of the safety factor is:

$$FS = \frac{(W \cos \alpha - ul) \tan \varphi'}{W \sin \alpha} \tag{1.14}$$

Substituting in Eq. 1.14 the value of u given by Eq. 1.13, and putting $c_u = \dfrac{\gamma_{sat} h}{4}$ (i.e. the value giving $FS = 1$ in terms of total stress), we obtain:

$$FS = \tan \varphi' \left(\cot \alpha - \frac{1}{\sin 2\alpha} \right) \frac{\sin \varphi' - 1}{\sin \varphi'} \tag{1.15}$$

Putting $\dfrac{dFS}{d\alpha} = 0$ we find that the minimum value of FS is obtained for $\alpha = 45° + \dfrac{\varphi'}{2}$.

Substituting this value in Eq. 1.15, we find that the expression again reduces to $FS = 1$. Both the total and effective methods of stability analysis agree in giving a factor of safety of unity but the position of the rupture surface is found to depend on the value of φ used in the analysis.

2 Review of pile types

2.1 Introduction

As mentioned before, in the last few decades the development in equipment and installation techniques and the pressure towards constructing in areas with poor subsoil properties have led to spectacular progress in the piling industry. At present the available piles range from micropiles with a diameter of 150 to 250 mm and load capacity of a few tens of tonnes to large diameter bored piles and large tubular steel piles of offshore structures with diameters up to two or three metres, length of many tens and sometimes over a hundred metres and load capacities of many hundreds or even thousands of tonnes.

Some of the situations in which pile foundations may be adopted are reported in Figure 2.1.

The most usual case is that of subsoil with loose or soft upper layers, not suitable for a shallow foundation. In this case the piles are used to transmit the load to

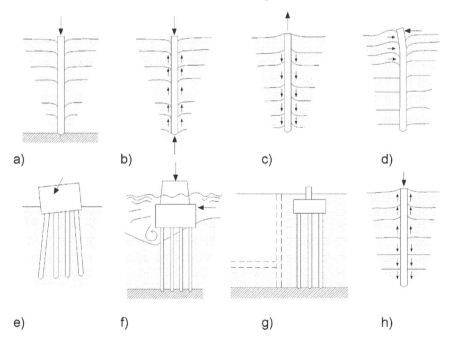

a)   b)   c)   d)

e)   f)   g)   h)

Figure 2.1 Typical pile applications (after Vesic 1977).

deeper competent layers, as for instance bedrock (point bearing piles, Figure 2.1a). If rock is not found within an acceptable depth, the load may be transmitted gradually by side shear (floating piles, Figure 2.1b). Side shear may resist upward loads (tension piles, Figure 2.1c). Horizontal forces may be resisted by vertical piles subjected to bending and shear (Figure 2.1d) or by pile groups including raking piles (Figure 2.1e).

Piles are generally associated with poor surface soils, but there are quite a number of different situations where they can be adopted advantageously. They are often used for bridge piers to keep the foundation below the maximum probable depth of scour (Figure 2.1f). In the scheme of Figure 2.1g an excavation adjacent to the foundation is foreseen, and piles transmitting the load below the depth of its bottom may be employed to prevent the possible adverse effect of the excavation. If swelling or collapsing soils are found near the soil surface, piles may be used to reach deeper soils, not affected by the seasonal water content variations and capable of resisting the upward or downward drag exerted by the shallow soils (Figure 2.1h).

Sometimes the adoption of a pile foundation is clearly dictated by the properties of the subsoil and the characteristics of the structure; in other instances it is chosen after a comparative analysis of possible alternative solutions.

For a shallow foundation and provided that correct constructional techniques are used, the disturbance to the soil is limited to a thin layer below the foundation depth (Figure 2.2). The mass of soil where a significant stress increase occurs or which would be involved in a possible shear failure, is practically unaffected by the constructional operations. The bearing capacity and settlement of the foundation can be thus predicted from knowledge of the mechanical characteristics of the undisturbed soil.

The conditions that govern the piled foundations are different. The soil in contact with the lateral pile surface, from which the pile derives its support by side friction, is deeply influenced by the pile installation. Similarly, the soil or rock beneath the toe of a pile is compacted or loosened to an extent that may affect significantly its end-bearing resistance. The behaviour of the pile, therefore, is markedly dependent on its installation technique; this factor substantially differentiates piles from shallow foundations.

The number of proprietary piles available on the market is very large and steadily increasing, because of the introduction of new techniques or new details in the exist-

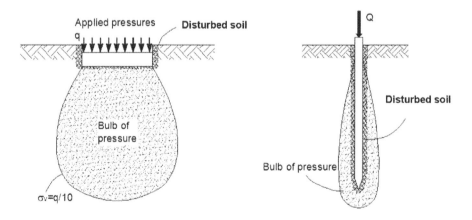

Figure 2.2 Difference between shallow foundations and deep foundations.

ing equipment and techniques; it is important, however, to have a clear idea of the principal pile installation techniques and their effects on the behaviour of the pile, since the choice of the pile type is an important part of the design. Furthermore, the effects of the various methods of pile installation on the behaviour of the pile can hardly be predicted by means of soil or rock mechanics theories; the best that can be done in practice is to apply empirical factors based on experience and on the results of field loading tests. The proper choice of such factors requires a clear insight into the effects of the various installation techniques.

Foundation piles may be classified following different criteria.

Referring to their *material*, piles may be made of wood, concrete or steel; concrete may be plane, reinforced, prestressed, cast in place or precast.

Referring to their *size*, pile can be subdivided into *small* (diameter $d \leq 250$ mm) *medium* (300 mm $\leq d \leq 600$ mm) and *large diameter* ($d \geq 800$ mm). Of course, these limits are largely conventional but they are of some practical use since design criteria are different for piles of different size.

The *installation technique*, however, is by far the most important differentiating factor among piles. The fundamental difference in installation techniques is that between *displacement* (or driven, pushed or screwed) piles and *replacement* (or bored or drilled) piles. In the former, there is no removal of soil, while in the latter a hole is previously bored, and the removed soil is replaced by concrete. There are, in addition, quite a number of intermediate pile types. Some of the most common types are listed in Table 2.1; some proprietary denominations have been reported when they are commonly used to designate a type.

Table 2.1 Principal types of piles

Replacement (bored, drilled) Small, medium or large diameter	Percussion or rotary bored, with/without casing, with/without bentonite mud, with/without enlarged base
	Continuous flight auger with grout or concrete injected from the shaft during extraction
Partial displacement Small, medium or large diameter	Small displacement (driven H sections; open ended pipes with soil inside removed, casing recovered after concreting)
	Continuous flight auger with large central stem, extracted with partial soil removal but allowing the placement of reinforcement before concreting (PressoDrill)
Displacement (inserted by driving, pushing or screwing) Small or medium diameter	Prefabricated	Wood Concrete: reinforced; centrifugated; prestressed
		Steel: closed end pipes; open ended pipes and H section with plugging
	Cast-in-place	Closed end concrete/steel pipes driven with mandrel, left in place and filled with concrete (Raymond, West)
		Recoverable casing and expanded base (Franki)
		Displacement screw piles (Atlas, Fundex, Omega, Discrepile,...)

2.2 Replacement piles

2.2.1 General

The installation of a replacement pile requires the removal of a cylindrical volume of soil from the ground, and filling the resulting void with concrete after having installed a reinforcement cage, if required. The essential factors of the installation are: (i) the drilling method; (ii) the temporary borehole wall support during excavation and concreting; and (iii) the concreting, in a hole possibly filled by water or drilling mud.

The *excavation method* may be percussion or rotary drilling, the latter being at present the most widespread. Continuous flight augering is also a form of rotary drilling.

The *temporary boring wall support* may be obtained by a recoverable casing or by the use of a suitable drilling mud.

The *concrete* may be simply *poured* from the top in dry holes, possibly with a temporary casing; it has to be *placed by tremie tube* if the hole is totally or partially filled with water or drilling mud.

2.2.2 Percussion boring

Percussion boring uses relatively simple equipment, typically a tripod and a friction capstan similar to but heavier than that used for SPT (Figure 2.3). Usual percussion piles have a diameter starting at 300 mm and rarely exceeding 600 mm; a steel casing is advanced with the excavation to support the soil. The cutting tool, called shell, is a heavy steel pipe with a cutting edge and a flap valve at the base; as it is dropped at the bottom of the hole, soil cuttings enter the tool; when it is raised, the valve closes and retains the soil inside. Water is usually added to soften the material. When the desired depth is reached, the reinforcement cage (if any) is placed and the pile concreted.

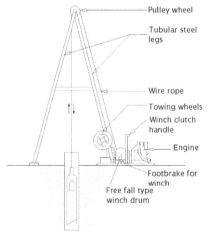

Figure 2.3 Percussion boring.

This kind of pile has severe limitation in size (small or medium diameter; depth hardly exceeding 20 m) and in the rate of installation. The main obstacle towards larger sizes and a more efficient process is the difficulty of advancing and recovering the steel casing. Casing is advanced by self-weight and driving. When a certain depth is exceeded, the single line pull of the winch is inadequate to extract the casing, and even a pulley system may be insufficient; hydraulic jacks are sometimes used.

Improvements of percussion bored piles included the use of grab shell and a semi-rotary driving action on the casing through a collar clamped to the same casing and operated by hydraulic rams originating from the crane. This system, known as the "Benoto" type, is represented in Figure 2.4; it is still in use to install piles with a diameter up to 1200 mm and a length up to 40 m.

2.2.3 *Rotary boring*

At present, the large majority of replacement piles of small, medium and large diameter are bored using rotary methods. Until a few years ago, drilling equipment was generally crane- or lorry-mounted with a power pack driving a ring gear with a rotary drilling table fixed to the crane base (Figure 2.5). This kind of equipment is still widespread; the lorry-mounted types are easier to transport between sites, while the crane-mounted ones are costly to transport, but are easier to move on uneven sites; the most powerful rigs are of the crane-mounted type. The drilling tool may be a core barrel, an auger or a bucket (Figure 2.6). It is operated from the ring gear via a square or keyed circular bar (the "kelly"), which may be telescopic for long piles. When filled with soil, the drilling tool is brought to the surface and emptied; the augered soil is removed from the auger by "spinning off" (Figure 2.7).

At present, the equipment of the new generation has both the rotating head and kelly hydraulically operated along a mast (Figure 2.8).

Modern equipment can drill holes up to 3.5 m diameter, and reach depth up to 50 m with a triple telescopic kelly or even more with an extension drill stem.

Figure 2.4 Percussion bored pile with oscillating collar (Benoto).

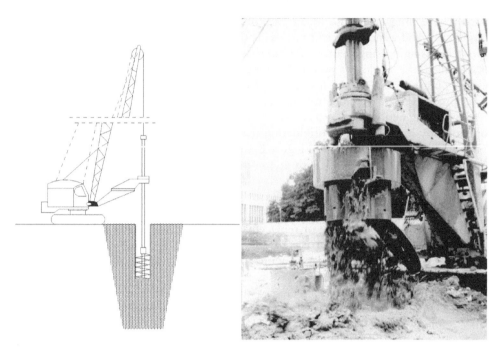

Figure 2.5 Rotary boring drilling equipment (track or crane mounted).

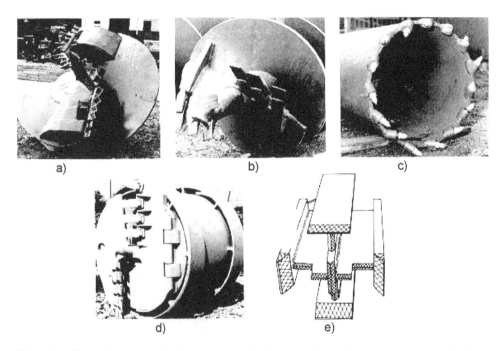

Figure 2.6 Tools for rotary boring: (a) standard auger; (b) rock auger; (c) coring bucket; (d) bentonite bucket; (e) chisel.

Figure 2.7 Soil removal from auger by spinning off.

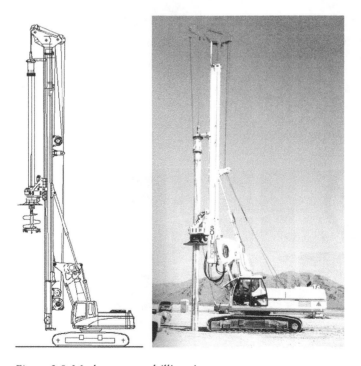

Figure 2.8 Modern rotary drilling rig.

Enlarged or underreamed bases can be cut by rotating a belling bucket within the previously drilled straight-sided shaft (Figure 2.9). Belling buckets can form enlargements of up to 6 m diameter, and require a shaft diameter of at least 800 mm to accommodate them; they are feasible only in relatively stiff cohesive soils.

When the bored pile has to be installed in a cohesive soil where the borehole wall is stable and there is no significant inflow of water, the hole may be excavated by the so-called *dry method*. Because the soil is self-supporting, there is no need to resort to excavation support techniques, except when the surface soil requires the use of a short casing (a few metres long) to avoid cave in. The reinforcement cage, if needed, is made with longitudinal bars and circular or spiral ties; the cage is inserted into the hole by a suitable winch and cable system, without hitting against the walls, which would cause soil to cave in. Both base and side spacers (Figure 2.10) are recommended. In a pile installed by the dry method, concrete is simply poured into the hole; it is important that the concrete falls without hitting the sides of the excavation and the reinforcing cage, as this might lead to concrete segregation and soil caving in and mixing with concrete. This is accomplished by a centring funnel and a drop

Figure 2.9 Tool for boring underreamed piles.

Figure 2.10 Spacers on the reinforcement cage.

chute guiding the concrete in free fall (Figure 2.11). Alternatively a flexible pressure hose is fed down to the bottom of the unlined hole and concrete is pumped while gradually raising the hose; this latter procedure is preferable.

In loose or weak soils, which are usual in the upper portion of a pile, or in water-bearing granular soils, some form of temporary support is necessary. This may be a temporary steel casing following the drilling tool, driven with a drop hammer or installed by a vibrator, and recovered by the rams operating the kelly bar by clamps gripping the casing, using a vibrator if needed. Modern hydraulic machines can install the casing by the same rotating head operating the drilling tools (Figure 2.12).

Figure 2.11 Drop chute.

Figure 2.12 Temporary steel casing following the drilling tool.

Concrete may be poured inside the temporary casing, provided it is not totally or partially filled with water. If there is inflow of water from the bottom, pumping it in an attempt to place the concrete in dry is not advisable, since the flow of water causes a progressive increase of the water content of the concrete, weakening it; a strong flow can even wash away the cement. In these cases water must be allowed to rise to its rest level into the hole, and concrete placed under water.

Where piles are required to penetrate considerable depth of unstable soils, the installation and recovery of a long casing is difficult and time consuming. In these cases a bentonite suspension of 5–6% by weight is used to support the borehole wall (*wet* or *slurry method*, Figure 2.13). This suspension has the property of forming a gel when allowed to stay static; when agitated by stirring or pumping, however, it has a liquid consistency. A positive head of bentonite slurry above the groundwater table of at least 1 m is desirable. In granular soils, because of this head and as its unit weight is slightly greater than that of water, the slurry penetrates the walls of the

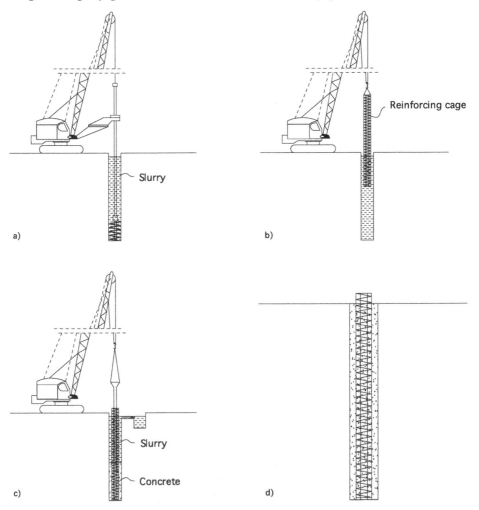

Figure 2.13 Slurry method: (a) drilling the hole; (b) insertion of the reinforcement; (c) concreting using a tremie pipe; (d) pile complete.

borehole where a "filter cake" rapidly builds up, preventing further flow and supporting the ground by the excess hydrostatic pressure. In fine-grained soils there is no penetration of the slurry but the excess hydrostatic pressure of the slurry prevents the collapse of the borehole wall anyway.

As a consequence of the environmental problems connected to the disposal of large volumes of mud, bentonite slurry is being gradually replaced by polymeric, biodegradable mud.

The placement of the concrete in a hole (either cased or uncased) completely or partially filled with water or drilling mud has to be performed by a tremie pipe (Figure 2.14). It has a diameter of 150 to 200 mm and is sealed at its lower end by a spherical rubber plug or a flap valve. The tremie pipe is lowered into the excavation; once it reaches the bottom of the hole, but without touching the soil, it is filled with a fairly fluid concrete, with a slump in excess of 150 to 175 mm. The plug at the end of the pipe is then displaced (or the valve opened) by the weight of the concrete column; the concrete exits the pipe and, being heavier than the water (or slurry), completely displaces the water (or slurry) from the hole. It is important to keep the lower end of the pipe always immersed in the concrete, in order to prevent any mixing of the concrete with the slurry; this requires that the tremie pipe be raised at the rate at which concrete accumulates into the hole. The bentonite slurry displaced by the concrete is allowed to settle in a holding tank to remove soil particles; it is then further cleaned, if necessary, with screens and centrifuges. The use of bentonite slurry causes disposal problems; the direct discharge into sewers or watercourses is not allowed, and the waste slurry has to be removed by tankers.

2.2.4 Continuous flight auger piles

A technique for installing replacement piles, which is gaining increasing popularity, is that of continuous flight auger or CFA piles, which in USA are also called augercast piles. They are installed by inserting into the soil a continuous flight auger by the combined action of a torque and an axial thrust (Figure 2.15). The auger has a

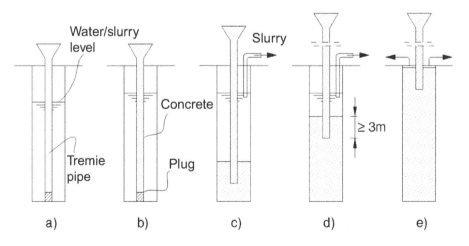

Figure 2.14 Tremie pipe.

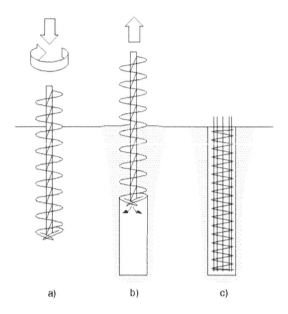

Figure 2.15 Continuous flight auger (CFA) piles: (a) insertion of the auger; (b) auger
  withdrawal and concrete casting; (c) insertion of the reinforcement cage into
  the fresh concrete.

hollow stem with an inner diameter of 60 to 100 mm provided with a seal at the
lower end. Once the desired depth is attained, a fairly fluid concrete is pumped down
the stem releasing the bottom seal whose function is to prevent the soil from entering. As concrete is pumped, the auger is pulled up without rotation removing the soil
within the spiral flights; the rate of raising the auger must be compatible with the
volume of concrete pumped. With this technique there is no need of temporary
support, since the wall of the borehole is continually supported by the spiral flights
and the soil within them, or by the concrete as it is pumped. Piles with a diameter up
to 1200 mm have been installed, but the typical range of diameters is between
400 mm and 800 mm.

CFA piles offer considerable advantages: vibration is minimal and noise is the
lowest for any pile of comparable size; productivity is high. The insertion of the
auger involves a slight displacement of the soil; for this reason, some authors classify
CFA piles among the partial displacement piles. Since the effects of the displacement
are substantially cancelled by the removal of the soil-filled auger, it is believed that
this pile type is included more properly among the replacement piles.

The drilling rigs are usually crawler mounted and quite tall, so that relatively long
piles can be installed (Figure 2.16). A concrete pump is required; modern rigs are
equipped with sensors to monitor and control the insertion parameters and the concrete placement.

If reinforcement is needed, it is introduced into the fresh concrete either pushing
the cage down manually, or vibrating or driving it. This is easily achieved for cages
up to 12–15 m in length; with a proper concrete mix and a suitably assembled cage,
depths up to 25 m have been reached.

Figure 2.16 Installing CFA piles.

In the cased CFA piles a counter-rotating casing tube is inserted and extracted together with the auger (Figure 2.17); this requires a double rotating head to operate simultaneously the casing and the auger. This kind of pile is frequently employed for the installation of secant piles diaphragms.

2.2.5 Micropiles

Micropiles are small diameter (150 to 250 mm) replacement piles that can be installed in almost any type of ground where piles are required.

Drilled micropiles were invented in Italy in the 1950s as a means to underpin sensitive historic buildings. At present, they compete for use with conventional larger diameter piles systems under many circumstances, and are especially suited in difficult ground conditions or with limited or difficult access, like inside buildings to be underpinned. A wide variety of drilling techniques can be employed: light auger, tricone, down-the-hole-hammer, casing with auger, percussion rod etc.

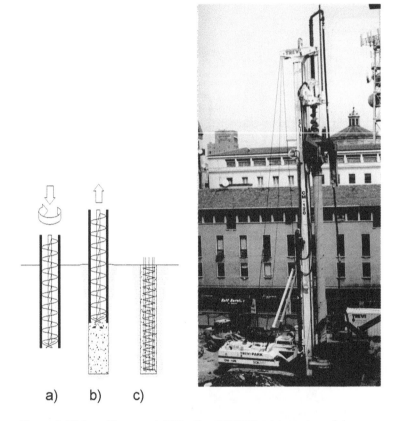

Figure 2.17 Installing cased CFA piles (TREVI) (a) insertion of the auger and of the case; (b) auger and case withdrawal and concrete casting; (c) insertion of the reinforcement cage into the fresh concrete.

The original *Palo Radice* or root pile (Figure 2.18), is a small diameter, cast in place reinforced drilled pile, generally designed to achieve a capacity in the range of 100 to 200 kN. Casing fitted with a tungsten bit is drilled into the ground to the full depth of the pile. The drilling fluid (either water or bentonite mud) is injected inside the casing and flows in the annular space between the casing and the soil carrying away soil cuttings. After drilling and placing the reinforcement (either a single bar or a small cage) sand-cement mortar is pumped via a tremie pipe from the bottom of the hole, displacing the drilling fluid. A pressure is applied to the grout during withdrawal of the casing. Pressure is typically in the range of 0.3 to 1 MPa, and is limited by the ability of the soil to maintain a grout-tight seal around the casing during its withdrawal.

In another micropile type presently available (Figure 2.19), a steel tube with no return valves (*tube à manchettes* or TAM), is placed in the hole and the annular space between the tube and the hole wall is filled from the bottom with a sand-cement mortar or neat cement grout. Fifteen to thirty minutes later, i.e. after setting but before hardening of the primary grout, a similar grout is injected at a pressure of 2 to 8 MPa using a double packer inside the TAM, so that specific horizons can be

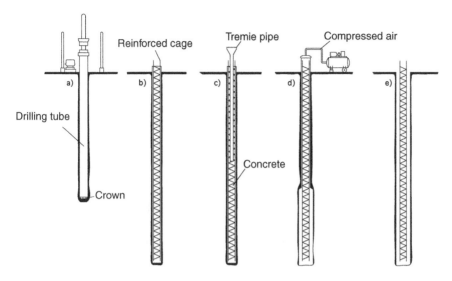

Figure 2.18 "Root" micropile (*Palo Radice*): (a) drilling; (b) reinforcement placement; (c) grouting with a tremie pipe; (d) extraction of the casing with pressure on the grout; (e) pile complete.

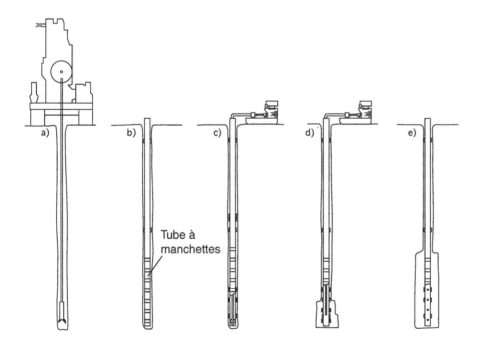

Figure 2.19 Micropile with TAM: (a) drilling; (b) insertion of TAM; (c) insertion of injection pile with packers; (d) injections; (e) micropile's shaft formed.

treated, if necessary, several times. To simplify the installation, sometimes all the valves are grouted simultaneously with a single packer placed just above the upper valve. At the end of the injection stage, the steel tube is filled with grout. The load is mainly resisted by the steel tube; capacity in the range of many hundreds to over 1000 kN may be achieved. The most usual dimensions and the maximum capacities are listed in Table 2.2.

2.3 Displacement piles

2.3.1 Driving equipment

The methods for installing displacement piles are jacking, vibration and driving. jacking and vibratory installation is comparatively rare.

The reaction needed to push a pile into the ground is equal to its bearing capacity, which can be very large; this made jacking suitable only for relatively small piles. Recently tubular steel piles of relatively large diameter have been installed by anchoring the jacking rig to previously installed piles (e.g. the "silent piler" of the Giken Company, Figure 2.20).

Vibratory driving is suitable for loose, saturated sand; it is routinely used to install sheet pilings and less frequently for steel H or pipe piles (Figure 2.21).

The most common method of installing displacement piles, however, is by driving the piles into the ground by blows from an impact hammer; piles installed in such a way are referred to as driven piles.

A variety of pile-driving hammers are available, differing from each other by the energy delivered in a single blow and by the number of blows per unit time. The existing types are: drop hammers; single-acting and double-acting vapour; compressed air or hydraulic hammers; and diesel hammers.

A drop hammer is a steel block weighing 10 to 50 kN (typically a half to two times the weight of the pile), which is raised through a pulley and dropped on the

Table 2.2 Micropiles with TAM

Nominal diameter (mm)	Diameter of the steel tube (mm)		Maximum allowable load (kN)	
	Outer	Inner	Tension	Compression
85	48.3	39.0	7	175
	51.0	35.0	135	235
100	60.3	44.3	165	300
		35.3	210	370
120	76.1	60.1	235	410
		51.1	280	510
145	82.5	66.5	255	520
		57.5	330	630
175	88.9	72.9	280	675
		63.9	360	795
200	101.6	85.6	320	845
		76.6	460	985

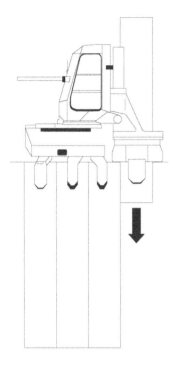

Figure 2.20 Giken silent pile.

Figure 2.21 Vibratory driving.

pile head. The height of fall can be varied in a wide range. The operator adjusts the energy delivered by varying the drop height; this implies the risk of damaging concrete piles under hard driving conditions through hard, strong soil.

In single-acting hammers (Figure 2.22a) the ram is connected to a piston, which is lifted in upstroke by steam, compressed air or hydraulic fluid and then allowed to fall under gravity in the downstroke. In double-acting hammers (Figure 2.22b) the pressure acts on the piston both in upstroke and downstroke, obtaining greater impact force and more impacts per unit time. The stroke is up to 1.5 m for single-acting, and smaller for double-acting hammers; the ram weight is in the range 20 to 150 kN.

A diesel hammer can also be single- or double-acting, and the pressure acting the ram is provided by the combustion of fuel injected before the stroke is completed. These hammers are smaller and lighter than other types, and very efficient.

The pile to be driven is positioned and kept aligned by a steel frame called a lead. The lead (Figure 2.23) has a height exceeding the pile length by some metres and is attached to a crane.

A number of elements are interposed between the hammer and the pile head, in order to avoid damaging it. They are referred to by various names, such as anvil, cushion, cap and helmet.

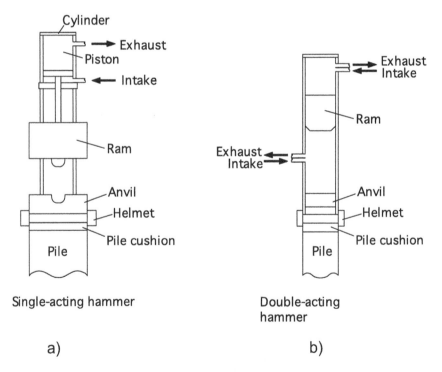

Figure 2.22 Single-action (a) and double-action (b) ram.

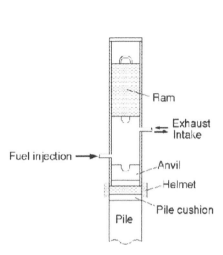 

Figure 2.23 Diesel hammer.

2.3.2 Prefabricated driven piles

Prefabricated piles are made of timber, concrete or steel.

Timber piles have an approximately circular cross section, with the diameter slightly decreasing from the top of the pile to the tip. The diameter ranges between 150 and 400 mm and the length between 6 and 20 m. The main problem with timber piles is their durability, since they are subject to decay when alternately exposed to wetting and drying, as around the water level in piles used in docks or waterfront structures or around the groundwater level in onshore foundations. Although different treatments are possible, none is 100% effective. When wooden piles are permanently under water, on the contrary, their duration may be quite long; many historical constructions, for instance in Venice (Italy) or in Stockholm (Sweden), are still satisfactorily resting on wooden piles.

The most common steel piles are pipe and H piles. Advantages of steel piles are their high resistance to driving and handling, their large flexural capacity and the ease with which they can be shortened or extended by cutting or welding. Their main shortcoming is susceptibility to corrosion in marine environment.

The cross sections commercially available for pipe and H piles are listed in Tables 2.3 and 2.4 respectively.

Prefabricated concrete piles are manufactured in plants and transported to the site; only for very large projects may temporary plants be installed at the site. Precast concrete piles may be in ordinary reinforced concrete or in pre-tensioned or post-tensioned prestressed concrete; ordinary reinforced concrete is preferred in almost all circumstances. The cross section is usually solid square, hexagonal or octagonal; for square piles, the maximum lengths to avoid excessive flexibility while handling and driving are listed in Table 2.5.

Detailed information on precast concrete piles is given by Tomlinson (1994).

Table 2.3 Commercially available pipe pile cross sections

External diameter (mm)	Thickness (mm)	Weight (kN/m)	Area (cm²)	Moment of inertia (cm⁴)
273	7.1	4.66	59.3	5245
	8.0	5.23	66.6	5852
	10.0	6.49	82.6	7154
	12.5	8.03	102	8697
323.9	7.1	5.55	70.7	8869
	8.0	6.23	79.4	9910
	10.0	7.74	98.6	12160
	12.5	9.60	122	14850
355.6	7.1	6.10	77.7	11810
	8.0	6.86	87.4	13200
	10.0	8.52	109	16220
	12.5	10.6	135	19850
406.4	7.1	6.99	89.1	17760
	8.0	7.86	100	19870
	10.0	9.78	125	24480
	12.5	12.1	155	30030
457	7.1	7.88	100	25400
	8.0	8.80	113	28450
	10.0	11.0	140	35090
	12.5	13.7	175	43140
508	7.1	8.77	112	35050
	8.0	9.86	126	39280
	10.0	12.3	156	48520
	12.5	15.3	195	59760
609.6	8.0	15.1	192	68410
	10.0	18.8	240	84680
	12.5	23.4	298	104540
	15.0	28.0	357	123910

Table 2.4 Typical H piles cross sections

Type	H (mm)	B (mm)	t (mm)	s (mm)	Area (cm²)	Weight (kN/m)	I_x (cm⁴)	I_y (cm⁴)
HEM 360	360	396	20.0	20.0	224	17.56	51770	20720
HEB 360	356	376	17.9	17.9	194	15.18	43880	15880
HEA 360	351	373	15.6	15.6	168	13.25	37480	13510
HLS 360	346	370	12.8	12.8	138	10.86	30200	10810
HEM 300	312	312	17.4	17.4	150	12.50	27030	8823
HEM 300	303	308	13.1	13.1	119	9.37	19630	6387
HEB 300	300	300	11.0	19.0	149	11.70	24790	8560
HEA 260	253	254	9.0	14.0	92.8	7.30	11200	3826
HLS 260	244	260	6.5	9.5	69.0	5.41	7981	2788
HEA 200	200	205	9	9	54.1	4.32	3888	1294

Table 2.5 Maximum practical length for square precast concrete piles

Side of the pile square section (mm)	Max length (m)
250	12
300	15
350	18
400	21
450	25

Large tubular concrete piles (Figure 2.24) have been produced in diameters typically of 900 to 1400 mm (Fleming *et al.* 1985); they are usually post-tensioned and often used in marine applications to sustain vertical loads up to 2 MN and over, combined with significant horizontal load.

Medium diameter hollow concrete piles, known in France as *pilotis* and in Italy as SCAC piles, are obtained by centrifugation; they are usually tapered, with a taper of 15 mm/m. The point is lined with sheet metal; when heavy driving is foreseen special steel points are adopted (Figure 2.25). The characteristics of the most commonly available piles are given in Table 2.6.

2.3.3 Cast in situ *driven piles*

Driven and cast in place piles are installed by driving a steel tube with the end closed. The pile is then filled with concrete, after placing a reinforcement cage if required, and the tube withdrawn. In other types thin steel shells or precast concrete shells are driven by an internal mandrel; after withdrawing the mandrel, reinforcement is placed if needed and concrete poured, leaving in place the shell.

The withdrawable tube types are installed by driving a steel casing closed at the tip by a loose steel plate. The tube is driven from the top by an impact hammer, and hence it is of heavy wall section. On reaching the required depth, a reinforcing cage is placed and a workable self-compacting concrete is poured in the tube through a

Figure 2.24 Large tubular prefabricated reinforced concrete pile.

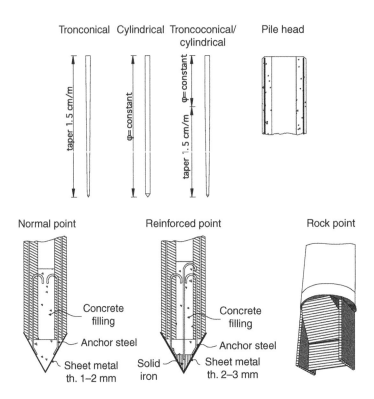

Figure 2.25 SCAC piles.

Note
th = thickness

hopper. The tube is then lifted by a hoist rope operated by the piling machine; it may be completely filled with concrete and lifted in a single stage, or lifted in stages. The necessity of recovering the casing limits the length of this kind of piles to 20 to 30 m; the diameter is in the range of medium diameter piles (300 to 600 mm).

An important variation to the above method is the Franki pile (Figure 2.26). A plug of gravel or dry concrete is introduced at the lower end of the casing, and an internal drop hammer acts on it. The friction which develops between the plug and the casing is sufficient for the blows to carry down the tube into the ground; during driving the tube is subjected to tensile stress instead of the compressive stress of other pile types, and thus it can be lighter because dynamic instability problems do not occur. When the required depth is reached, the tube is restrained from further penetration and batches of dry concrete are hammered out to form a bulb or expanded base of the pile. The reinforcing cage is then inserted, and dry concrete is placed in batches and compacted by the internal hammer while the casing is pulled out in stages. The operation of concreting and compacting in stages is slow and cumbersome; accordingly a complete filling with workable concrete and recovering the casing in a single stage are gradually replacing it. Franki piles are also limited to the range of medium diameters ($300\,\text{mm} \leq d \leq 600\,\text{mm}$) and to lengths of 20 to 30 m. A rig to install Franki piles is represented in Figure 2.27.

Table 2.6 Centrifugated concrete hollow tapered piles (SCAC)

L (m)	Top diameter (mm)	Thickness (mm)	Weight (kN)	Reinforcement	Maximum load (kN)	
					Top	Tip
Tip diameter 220 mm; tip thickness 55 mm						
6	310	70	6.1	6 Ø 7	330	300
8	340	75	7.5	8 Ø 7	440	300
10	370	80	12.5	11 Ø 7	610	310
12	400	85	16.5	11 Ø 8	800	320
Tip diameter 240 mm, tip thickness 60 mm						
8	360	85	10.5	8 Ø 7	440	320
		95	11.0	8 Ø 8	720	390
10	390	85	14.5	11 Ø 7	610	380
		95	15.0	12 Ø 8	870	400
12	420	90	18.5	11 Ø 8	800	400
		100	20.0	12 Ø 9	1100	420
14	450	95	24.0	12 Ø 9	1100	420
		105	25.0	12 Ø 10	1320	450
16	480	100	30.0	14 Ø 10	1400	470
		110	32.0	14 Ø 12	1560	520
Tip diameter 260 mm, tip thickness 70 mm						
12	440	100	23.0	12 Ø 9	1100	530
14	470	105	29.0	14 Ø 9	1280	540
16	500	110	36.5	14 Ø 12	1630	600
18	530	115	44.5	16 Ø 12	1820	640
20	560	120	53.5	20 Ø 12	2060	710

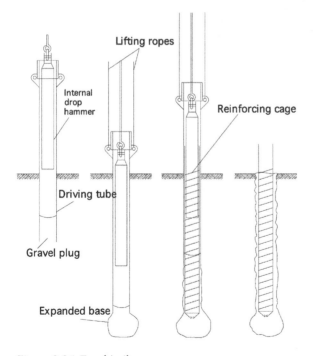

Figure 2.26 Franki piles.

Figure 2.27 Rig for installing Franki piles.

Piles employing a permanent metal shell generally consist of a thin corrugated steel tube locked to a steel bottom plate. The lining tube and the shoe are driven by a mandrel which locks into the corrugation (Figure 2.28); once reached the required depth the mandrel is collapsed and withdrawn and the permanent liner is filled with concrete and any necessary reinforcing steel. The Raymond step taper pile is made by helically corrugated light steel shells assembled on a mandrel, generating a tapered steel shell (Figure 2.29). The West shell pile (Figure 2.30) incorporates precast concrete shell units, which are threaded onto an internal driving mandrel with a detachable concrete conical shoe at the end.

2.3.4 Displacement screw piles

Displacement screw piles are installed with special augers designed to displace the soil during installation; some of these tools are shown in Figure 2.31.

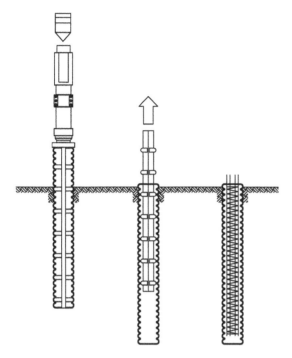

Figure 2.28 Lacor piles.

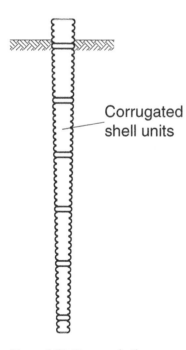

Corrugated
shell units

Figure 2.29 Raymond piles.

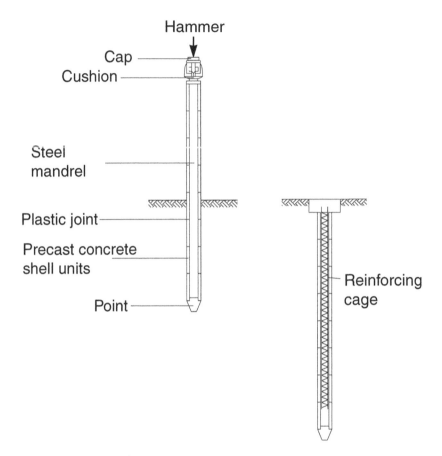

Figure 2.30 WEST piles.

   The tool contains typically a soil displacement body, a helical segment, and a sacrificial tip. The tool is connected to a stem whose diameter may be equal or smaller than the nominal diameter of the pile. At the end of the penetration, the tip is released and concrete or grout is injected through the hollow stem while the tool and the stem are withdrawn from the soil. Depending on the diameter of the hollow stem, reinforcement may be inserted either before or after the concrete placement.
   Typical diameters are in the range 0.3 to 0.6 m with lengths up to 24 m.

2.4 Partial displacement piles

2.4.1 *Driven H or tubular piles*

The characteristics of H and tube steel piles have been reported in §2.3.2. The tubular piles are driven open-ended; if plugging occurs, the soil inside the tube is removed by an auger. Once the desired depth is reached, a reinforcing cage (if required) is placed. If the tube is dry, concrete may be poured from the surface with a centring funnel; if the tube is full of water, concrete is placed from the bottom by a tremie pipe.

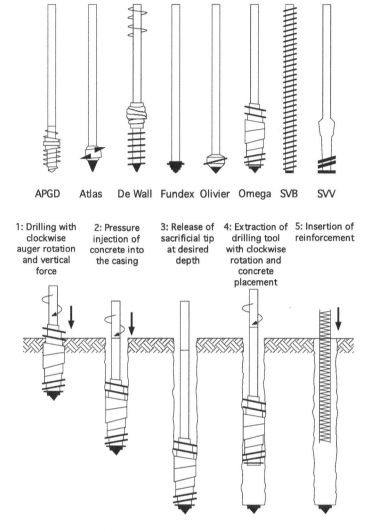

Figure 2.31 Displacement screw piles.

2.4.2 Large stem auger piles (PressoDrill, SVB)

These piles represent a modification of CFA piles, in which the diameter of the central hollow stem is substantially increased (and the width of the outer spiral correspondingly decreased) in order to make possible the installation of a reinforcing cage before concreting. With reference to Figure 2.32, if d is the diameter of the flight auger (corresponding to the nominal diameter of the pile) and d_o that of the central hollow stem, the ratio d_o/d is in the range 0.1 to 0.15 for usual CFA piles and in the range 0.7 to 0.9 for this pile type.

As for CFA piles, the auger is inserted into the soil by the combined action of an axial thrust and a torque; the penetration involves a displacement but also some removal of the soil. Neglecting the volume of the auger blade, the volume

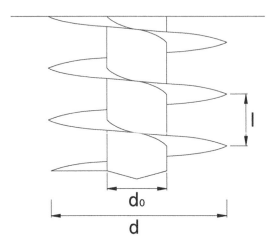

Figure 2.32 Analysis of the installation of a large stem auger pile.

V_d of soil displaced by the auger penetration in the time interval Δt may be expressed as:

$$V_d = \frac{\pi d_o^2}{4} v\Delta t$$

where v is the velocity of penetration; the volume V_r of soil removed by the screw in the same time interval is:

$$V_r = \frac{\pi(d_o^2 - d^2)}{4}(nl - v)\Delta t$$

If $v = nl$ (i.e. if the vertical travel of the auger during one revolution is equal to the pitch l of the screw), $V_r = 0$ and the auger penetrates the soil as a screw without removing any material. At the other extreme, if $v = 0$, the auger acts as an Archimedean pump removing a volume of soil V_{rmax} obtained by the above expression with $v = 0$.

The insertion of the screw produces a net compression effect provided that the displaced volume exceeds that removed; to this aim, the following condition must be satisfied:

$$v \geq nl\left(1 - \frac{d_o^2}{d^2}\right)$$

In practice, v is always smaller than nl; nevertheless the net effect on the soil may still be compression.

Viggiani (1993) has shown that, if the torque needed is always available, the axial thrust needed to get a penetration without decompression of the soil increases with increasing depth, reaches a maximum at a depth z in the range 5 to 10 times the diameter d, and eventually vanishes at a depth in the range 10 to 20 times d. The maximum thrust required increases with increasing the strength of the soil and the ratio d_o/d. If the equipment has not a sufficient thrust capacity, the penetration slows down and some soil removal during penetration cannot be avoided. As a consequence, the soil surrounding the pile loosens and the penetration becomes possible; the bearing capacity of the pile, however, decreases.

2.5 Advantages and shortcomings of the different pile types

The principal advantages and disadvantages of the various types of piles described above can be summarized as follows.

Displacement piles are in general suited in cohesionless soils, since their installation increases the horizontal stress and decreases the porosity in the soil surrounding the pile, thus improving its strength and stiffness and increasing the stiffness and bearing capacity of the pile. On the contrary, they have to be adopted with caution in cohesive soils, since their installation remoulds the surrounding soils and induces high excess pore pressure.

Prefabricated driven piles are not liable to squeezing or necking, and they can be inspected for soundness before driving. Their installation is not affected by groundwater, and their projection above ground level may be advantageous in structures in shallow water, such as jetties, harbours etc. On the side of shortcomings, they are limited in length and diameter, and cannot readily be varied in length to suit varying depths of bearing layers. They cannot be driven in soils including cobbles or thin very dense or cemented layers. The noise and vibrations due to driving may be unacceptable, especially in an urban environment; the displacement of soil due to driving may lift adjacent piles or damage adjacent structures. They cannot be driven in conditions of low headroom.

Cast *in situ* displacement piles can be easily adjusted to suit varying depths of bearing layers; concreting is generally possible excluding ground water; the formation of an enlarged base is possible in some types. Driving with an internal drop hammer, noise and vibrations may be reduced.

On the other hand, the concrete in the shaft is liable to be defective in soft squeezing soils or in condition of artesian water flow, when a recoverable tube is used. The length of some types may be limited by the capacity of the rig to pull out the tube casing. Displacement screw piles have limits in diameter and length; the reinforcement cage has generally to be inserted into the fresh concrete.

Replacement piles, with the exception of CFA, generally exhibit relatively low resistance in cohesionless soils due to the inevitable loosening of the soil below the base and surrounding the shaft.

The range of length and diameters available is very broad, from micropiles to large diameter shafts; the length can readily be varied to suit variations in soil profile; any kind of soil, including boulders and rock layers, may be drilled by choosing a suitable technique. Replacement piles can be installed without noise or vibration, and also in conditions of low headroom.

As for all the cast *in situ* types, the concrete is liable to squeezing or necking in soft soils; special techniques are needed for concreting in water- or mud-filled holes.

3   Design issues

3.1 The steps of design

The design of any foundation, including pile foundations, develops through a number of logical steps, which are:

1   Collection of geological evidence and any other available information on the subsoil; planning, execution and interpretation of site and laboratory investigations on the subsoil; development of a *geotechnical characterization* of the subsoil.
2   Determination of the magnitude, nature and distribution of the loads exerted by the structure on the foundation. Loads may be: *permanent* (dead weight of the structure and of any material, machinery or finishing permanently fixed to it); *live*, distinguishing between the loads acting for long periods of time (materials in warehouses or silos, liquids in tanks, earth pressure on retaining structures) and for short intervals (wind or snow; crowd in a building; trucks on a bridge); *dynamic*, such as seismic actions or vibrations by machinery or traffic; *cyclic*, such as wave action on coastal and maritime structures.
3   Choice of the type of foundation; in some cases the choice is obvious, in other a preliminary analysis of alternative solutions may be needed. Once the decision of adopting piles has been taken, selection of the pile type is made by taking into account the nature of the soils but also the local market and possible constraints at the construction site (access for equipment, effect of vibration on the environment).
4   Determination of the bearing capacity of the single pile and the overall pile foundation (pile group, plus the contribution of the structural element connecting pile heads, if this is the case). Choice of a tentative service load, obtained assuming a suitable margin of safety against collapse (bearing capacity failure).
5   Prediction of the total and differential settlement of the foundation; assessing its admissibility taking into account the static and functional characteristic of the structure. If the admissibility criterion is not satisfied, decrease of the service load per pile or change of the foundation scheme.
6   Evaluation of the stress in the foundation structure and structural design.
7   Definition of the installation techniques and preparation of the technical specifications.
8   Evaluation of the cost, also to assist in the choice between possible alternative solutions.

In evaluating safety against a bearing capacity failure, the dead loads plus all possible loads have to be taken into account. The most unfavourable combinations of different live loads must be the same used for the design of the vertical bearing elements of the ground floor of the structure.

If the subsoil is composed by coarse-grained, free-draining soils (sand, gravel), the above load has to be considered in evaluating the settlement also. If, on the contrary, the subsoil is composed by fine-grained soils (silt, clay), the load to be considered is the dead load plus a fraction of the live loads depending on the frequency and duration of their action. In the lack of better, the fraction of live loads may be assumed equal to 50%.

Items 4 and 5 + 6, which in some ways dominate the design calculations, are the result of a widespread rethinking of the whole civil engineering design process taking place in the post-war boom of the construction industry and leading to the introduction of two particular ratios (Institution of Structural Engineers 1955):

• the ratio of the ultimate load to the appropriate working load, known as the ultimate load factor or overall safety factor against collapse; and
• the ratio of the limiting load, causing excessive elastic deflections or local defects such as cracks etc. to the appropriate working load, known as limiting load factor.

Brinch Hansen (1956) claims that: "In the design of any structure two separate analyses should in principle be made: one for determining the safety against failure and another for determining the deformations under actual working condition."

Some suggestions from the available literature or daily practice are summarized in the following.

3.2 Overall factor of safety

The overall factor of safety approach has been customary in geotechnical engineering until recently, and is still widely adopted in practice.

The approach calls for the determination of an ultimate bearing capacity Q_{ult} of the foundation, employing the best available estimate of the soil properties. The design requires that:

$$\frac{Q_{ult}}{FS} \geq \sum P_i$$

where *FS* is the global factor of safety and P_i are the applied loads.

Standards and Codes of Practice, based on experience and precedents, suggest or prescribe values of FS; for pile foundations, values between 2 and 3.5 have been used (Poulos *et al.* 2001). Some suggested values of *FS* for pile foundations are listed in Table 3.1.

As a matter of fact, neither the bearing capacity of a foundation nor the applied load are deterministic quantities. Uncertainties in the design analyses and the subsoil characterization, and the random variations of soil properties and pile installation effects affect the bearing capacity, while uncertainties in the acting load, geometry of the foundation, behaviour and characteristics of the superstructure influence the

Table 3.1 Suggested values of the overall factor of safety *FS* for pile foundations

Characteristics of the structure	Homogeneous subsoil; exhaustive site and laboratory investigations	Unsatisfactory subsoil characterization because of either heterogeneous soil or poor investigations
Maximum design loads occur frequently; the consequences of a collapse would be catastrophic (e.g. chemical or nuclear plant)	3	3.5
Maximum design loads occur rarely; the consequences of a collapse would be heavy (e.g. road bridges)	2.5	3
Maximum design loads are very improbable (e.g. residential buildings)	2	2.5

design value of working load. Some attempts have been made to relate safety factors to statistical parameters of the ground and the foundation type, and to the admissible risk of failure. The reliability-based design is not yet sufficiently developed for general practical application, but the limit state approach with partial factors of safety, at present almost universally adopted, may be seen as a simplified form of reliability-based design.

3.3 Limit state design

3.3.1 Introduction

A limit state is a set of conditions to be avoided; it may be either an ultimate or a serviceability limit state.

According to Ovesen (2002), whenever a geotechnical structure (or part of it) fails to satisfy one of its performance criteria, it is said to have reached a "limit state". In a code based on "the limit state method" (such as, for instance, Eurocodes) each limit is considered separately in the design and its occurrence is either eliminated or shown to be sufficiently improbable.

Efforts have been made in order to refine principles, criteria and numerical values of the partial factors for any of the considered limit states. Furthermore, in view of a globalization even in engineering profession, need of harmonization of the various design codes in the world rapidly raised up (IWS 2002).

The ultimate limit state is associated with the concept of failure and hence of danger; in foundation engineering it results from a bearing capacity failure of the foundation, but also from excessive differential settlement leading to partial or full structural collapse.

A structure reaches a serviceability limit state when it does not perform its intended function. In foundation engineering the serviceability limit state is associated with settlement. Excessive settlement, even if uniform, might lead to failure of connections to utility lines or access problems. Excessive differential settlement produces cracking of plaster and panel walls, jamming of windows and doors and similar architectural damages.

Prevention of serviceability limit state usually precludes ultimate limit state as well, but regulations and practice require the engineer to independently check each possible limit state.

It is pointed out that independent checks for excessive foundation settlement have long been performed in geotechnical engineering practice, even when adopting the traditional global factor of safety approach; the formal use of the expression "serviceability limit state", however, is relatively recent.

3.3.2 Ultimate Limit State – ULS

The check for ultimate limit state may be performed following two slightly different approaches: the Load and Resistance Factor Design (LRFD) or the Partial Factor (PF) approach.

The former is sometimes referred to as the American Approach, being widely adopted in the USA, while the latter is known as the European Approach, because of the considerable extent of its application in that continent.

The LRFD or American approach is expressed as follows:

$$RF \cdot Q_{ult} \geq \Sigma_i (LF)_i \cdot Q_i \qquad (3.1)$$

where $RF \leq 1$ is the resistance factor, Q_{ult} the nominal bearing capacity, $(LF)_i \geq 1$ are the load factors and Q_i the nominal loads. The subscript i refers to the load types (dead load, live load etc.) and to the related factors; the term "nominal" (or "characteristic") means that the loads and the bearing capacity are evaluated following certain guidelines. The nominal bearing capacity is that computed using the best estimate of the shear strength properties of the soil; accordingly, it is equal to the value employed in the global factor of safety approach.

Eq. 3.1 states that the factored load should not exceed the factored resistance.

The European approach requires that:

$$\Sigma_i (LF)_i \cdot Q_i \leq Q'_{ult} \qquad (3.2)$$

where Q'_{ult} is the bearing capacity calculated using soil strength parameters obtained by reducing the characteristic strength values of the soil with partial factors of safety. In other words, the factored load should not exceed the resistance as computed by factored shear strength parameters.

The two approaches are compared in Figure 3.1 (Becker 1997).

3.3.3 Serviceability limit state – SLS

The design against serviceability limit states requires that the overall and differential displacement of a foundation be kept below the corresponding admissible values. Referring to Figure 3.2, the quantities to be considered are:

- the overall settlement w;
- the differential settlement δ_{AB} between two points A and B;

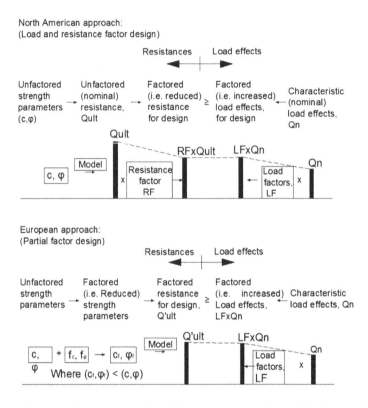

Figure 3.1 European vs. North American approach to ULS design (after Becker 1997).

- the tilt α;
- the angular distortion or relative rotation β_{AB} between two points A and B, equal to the differential settlement divided by the distance L_{AB} between the two points $(\beta_{AB} = \delta_{AB}/L_{AB})$;
- the relative deflection Δ_{AB} of wall and panels between two points A and B; and
- the deflection ratio equal to the relative deflection divided by the distance between the two points $(\eta_{AB} = \Delta_{AB}/L_{AB})$.

Data on the allowable values of the above quantities have been collected by a number of sources. Poulos *et al.* (2001) have distilled from the previous indications the values listed in Table 3.2. The procedure for the prediction of the differential displacements of a pile foundation are reviewed in §5.4.

The use of a single quantity, such as angular distortion or deflection ratio, to assess building damage excludes many important factors, related to the building (flexural and shear stiffness, geometrical configuration), to the subsoil (coarse-grained or fine-grained soils, and related differences in the rate of occurrence of settlement) and to the nature of ground movement profile (e.g. sagging or hogging). Boscardin and Cording (1989) pointed out the importance of horizontal strain, and derived the relation shown in Figure 3.3 between the degree of damage, the

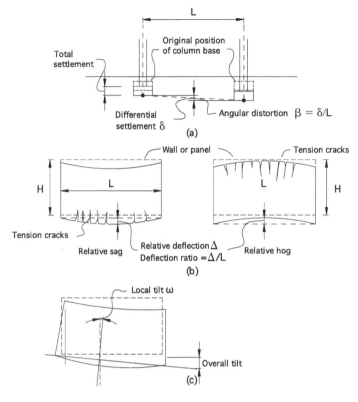

Figure 3.2 Quantities defining the deformation of a foundation (after Burland and Wroth 1974).

Table 3.2 Admissible values of settlement and distortion of structures

Type of structure	Type of damage/ concern	Quantity to be considered	Limiting value
Framed building and reinforced load bearing walls	Structural damage	Angular distortion	1/150–1/250
	Cracking in walls and partitions	Angular distortion	1/500 (1/100–1/1400 for end bays)
	Visual appearance	Tilt	1/300
	Connection to services	Total settlement	50–75 mm (sand) 75–135 mm (clay)
Tall buildings	Operation of lifts and elevators	Tilt after lift installation	1/1200–1/2000
Structures with unreinforced load bearing walls	Cracking by sagging	Deflection ratio	1/2500 (L/H = 1) 1/1250 (L/H = 5)
	Cracking by hogging	Deflection ratio	1/5000 (L/H = 1) 1/2500 (L/H = 5)
Bridges – general	Ride quality	Total settlement	100 mm
	Structural distress	Total settlement	63 mm
	Function	Horizontal movement	38 mm
Bridges – multiple span	Structural damage	Angular distortion	1/250
Bridges – single span	Structural damage	Angular distortion	1/200

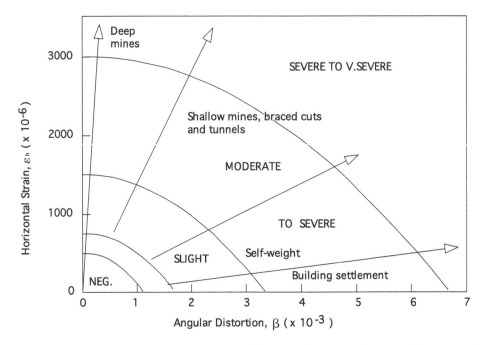

Figure 3.3 Degree of damage as a function of horizontal strain and angular distortion (after Borscardin and Cording 1989).

horizontal strain and the angular distortion. It may be seen that the larger the horizontal strain, the less is the tolerable distortion before some form of damage occurs. Such consideration may be important when assessing potential damage arising from tunnelling operations, open excavations and adjacent constructions.

Part II

Present practice of piled foundations design under vertical loads

4 Bearing capacity under vertical load

4.1 Introduction

Piles are rarely used as single piles (this practice is being modified because of the present availability of large diameter, large capacity piles), but in groups ranging from a few units for piled footings of single, heavily loaded columns to large piled rafts resting on many hundreds of piles.

Design practice, however, is based on the evaluation of the behaviour (bearing capacity, settlement under working load) of a single pile; the behaviour of the pile groups is then obtained from that of the single pile.

In order to minimize the cost of the cap, the piles in a group are usually kept as close to each other as possible, the only limit being the possible damage to the already installed piles. The usual spacing ranges between $2.5d$ and $3.5d$ axis to axis.

4.2 Definition of bearing capacity

Vertical load is by far the most common and most relevant load condition for pile foundations, as for any other kind of foundation.

The load Q applied to the pile head is transmitted to the soil partly by normal stress p at the pile base, and partly by shear stress τ at the lateral pile–soil interface. Vertical equilibrium requires that (Figure 4.1):

$$Q = \frac{\pi d^2}{4} p + \pi d \int_0^L \tau \cdot dz = P + S$$

where d is the diameter and L the length of the pile; P is total point load and S the total side resistance (or side friction or shaft resistance).

If a vertical axial load Q is applied to a pile, a settlement w of the pile head is observed. In Figure 4.2 a typical load–settlement response of a pile is reported; the Q–w curve is non-linear from the very beginning, and eventually merges into a vertical asymptote corresponding to the load required for pile penetration into the soil.

Load tests on instrumented piles have shown that in the first stage of loading, the load is resisted by shear stress along the pile shaft. As the load is increased, the full shaft resistance is mobilized and load begins to be transferred to the pile base. Since the mechanism of mobilization of the shaft resistance is that of sliding of two

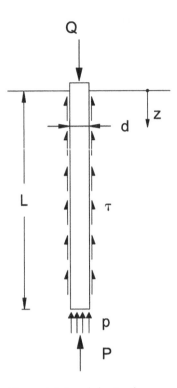

Figure 4.1 Load sharing between the shaft and the base for a pile under vertical load.

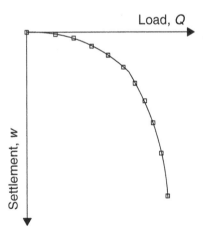

Figure 4.2 Load–settlement response of the head of a pile under axial load.

bodies on a contact surface, the displacement needed to mobilize the maximum shaft resistance is almost independent of the pile diameter and relatively small (a few tens of millimetres). On the contrary, the base resistance is mobilized only at large pile displacement, proportional to the pile diameter. To explain this finding, reference may be made to Figure 4.3, where an ideal load–settlement curve of the pile base is

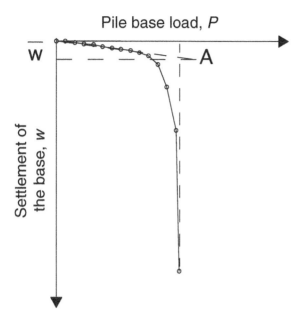

Figure 4.3 Load–settlement response for the pile base.

shown. If the pile is in a clay soil and the load is applied under undrained conditions, the initial tangent to the load–settlement curve may be expressed as:

$$w = \frac{p \cdot d}{E_u} I_w$$

where E_u is the undrained modulus of the clay and I_w is an influence factor which, for a deep circular loaded area, is around ½. The settlement $\bar{w}$ corresponding to a linear behaviour up to failure (point A in Figure 4.3) may be evaluated taking into account that the value of p at failure is $9c_u$ and a typical value of E_u is in the range 200 to 400 times c_u. One obtains:

$$\bar{w} = \frac{1}{2} \frac{9c_u d}{(200 \div 400)c_u} = (0.01 \div 0.02)d$$

that is, a value proportional to the pile diameter.

This explains why the settlement needed to mobilize point resistance increases with the pile diameter. As a matter of fact, according to the experimental evidence the settlement at ultimate failure ranges from $0.1d$ (displacement piles) to $0.25d$ (replacement piles).

Such large displacements are rarely attained in load tests and are of little practical significance. It is generally accepted that the bearing capacity of a pile be conventionally defined as the load corresponding to a settlement equal to 10% of the pile diameter.

Randolph (2003) discusses the factors controlling the bearing capacity of piles in clay and sand, concluding that while the "science" provides the framework within which they are to be considered, design calculation still rely on empirical correlations.

At this time, one cannot but agree with Poulos *et al.* (2001) that it is very difficult to recommend any single approach as being the more appropriate for estimating the axial bearing capacity of single piles. Given the very nature of the problem, the most reasonable approach seems that of going on developing regional design methods combining the local experience of both piling contractors and designers.

A review of the present practice is presented in the following paragraphs.

It is almost universally accepted that the bearing capacity Q_{ult} of a pile can be estimated by summation of the ultimate shaft capacity S_{ult} and the ultimate base capacity P_{ult}. The weight W of the pile is subtracted from the compressive load and added to the uplift capacity. In turn, S_{ult} and P_{ult} may be expressed as a function of the ultimate unit shaft and base resistance s and p. Thus, for compression:

$$Q_{ult} = \pi d \int_0^L sdz + \frac{\pi d^2}{4} p - W$$

and, for uplift

$$Q_{ult} = \pi d \int_0^L sdz + \frac{\pi d^2}{4} p + W$$

Of course, the ultimate values of p and s are not necessarily the same in compression and uplift; in particular, the uplift resistance of the pile base is usually disregarded.

There are two main approaches to the calculation of the pile capacity: (i) from fundamental soil properties, and (ii) from *in situ* tests results.

4.3 Bearing capacity from fundamental soil properties

4.3.1 *Medium diameter piles; base resistance*

The base resistance may be obtained, similarly to shallow foundations, by an expression of the following type:

$$p = N_q \sigma_{vL} + N_c c + N_\gamma \gamma \frac{d}{2} \tag{4.1}$$

where σ_{vL} represents the overburden stress at the depth L of the pile base and N_q, N_c, N_γ are bearing capacity coefficients, function of the friction angle φ and possibly of the ratio L/d.

N_q and N_γ are of the same magnitude, and L typically 50 to 100 times greater than $d/2$; therefore $\sigma_{vL} = \gamma L$ is much larger than $\gamma d/2$ and hence the third term of the expression may be neglected. Since the pile section is generally isometric (either circular or square) the bearing capacity coefficients N_q and N_c refer to an axially symmetric condition. By Caquot's theorem of corresponding states, the coefficient N_c may be obtained by the relation:

$$N_c = \left(N_q - 1\right) ctg\varphi$$

In undrained conditions (clay soils), Eq. (4.1) may be applied in terms of total stress. With $c = c_u$ and $\varphi = 0$, it has been found that $N_q = 1$ and $N_c = 9$, and hence:

$$p = \sigma_{vL} + 9c_u \qquad (4.2)$$

where σ_{vl} is the total overburden stress at the depth L of the pile base.

In drained conditions (sandy soils), the analysis is carried out in terms of effective stress; in general the effective cohesion is negligible and one obtains:

$$p = N_q \sigma'_{vL} \qquad (4.3)$$

where σ'_{vL} is the effective overburden pressure at the depth L of the pile base.

The relation between N_q and the friction angle φ, and possibly the soil compressibility and the ratio L/d, is given by a proper theory.

Figure 4.4 reports the hypotheses under which different theories have been developed. Initially, the classic rigid-perfectly plastic solutions developed for shallow foundations in plane strain have been adopted (Figure 4.4a), introducing corrective coefficients for the circular shape and for the depth. The same material model, but a different slip line field has been adopted by others (Figure 4.4b). Berezantsev *et al.* (1961) (Figure 4.4c) claim that the vertical stress at the depth of the pile point is less than overburden by a sort of silo effect; in their solution, accordingly, the values of N_q are function of the ratio L/d, in addition to φ. Finally, using an elasto plastic material model and adapting the theory of the expansion of a spherical cavity, Vesic (1977) (Figure 4.4d) produces values of N_q function of a reduced rigidity index $I_{RR} = I_R/(1 + I_R\delta)$, where $I_R = \dfrac{G}{c + \sigma \cdot tg\varphi}$, G is the shear modulus of the soil and δ is the average volume strain below the pile base.

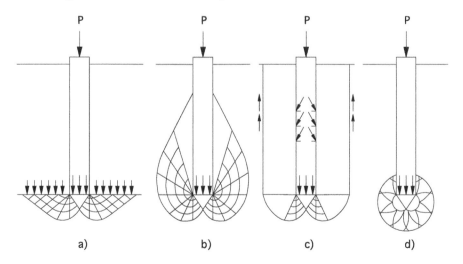

Figure 4.4 Different proposals for the slip line field below the pile base: (a) Prandtl (1921); Caquot (1934); (b) De Beer (1945); Meyerhof (1953); (c) Berezantsev *et al.* (1961); (d) Bishop *et al.* (1945); Skempton *et al.* (1953); Vesic (1964, 1977).

The different theories give markedly different values of N_q (Figure 4.5).

Basic experimental investigations carried out by Kérisel (1961) and Vesic (1964) revealed that the unit point resistance p does not increase linearly with depth as predicted by Eq. 4.3. On the contrary, below a certain critical depth, p keeps practically constant with depth. Vesic claims that this is an effect of φ' and I_{RR} decreasing with increasing σ' (Table 4.1).

From a practical viewpoint, adopting the Berezantsev *et al.* (1961) theory is suggested; the corresponding values of N_q are reported in Figure 4.6.

Finally, it is to be considered that the friction angle of the soil surrounding the pile is modified by the pile installation; because of the large gradient of N_q with φ, this is a further significant problem in the prediction of the base resistance of piles. If φ is the friction angle of the soil before the installation of the pile, to account for this effect Kishida (1967) suggests to assume, in the computation of bearing capacity, a value φ_m modified as follows:

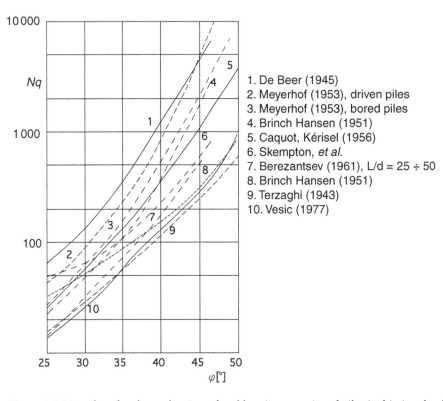

Figure 4.5 Nq values for the evaluation of end bearing capacity of piles in frictional soils.

Table 4.1 Properties of Chattahochee River sand at DR = 80%

Effective stress σ' (kPa)	φ' (degrees)	I_{RR}	N_q (Vesic)	$N_q \sigma'$ (kPa)
70	45	122	280	19 600
850	32.5	12	23	19 550

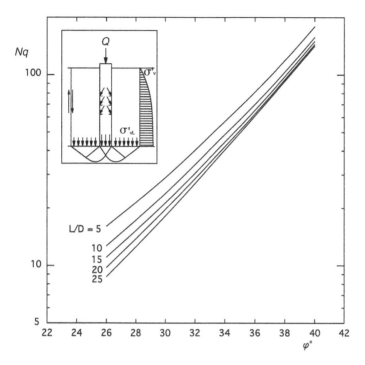

Figure 4.6 Values of the bearing capacity coefficient *Nq* according to Berezantsev *et al.*
(1961).

- $\varphi_m = \dfrac{\varphi + 40°}{2}$ for displacement piles
- $\varphi_m = \varphi - 3°$ for replacement piles

4.3.2 Medium diameter piles; shaft resistance (side friction, lateral resistance)

The general expression of the unit shaft resistance is:

$$s = a + \sigma_h \mu$$

where *a* is a cohesive term, σ_h is the horizontal normal stress acting at the interface
between the pile and the soil at the depth of the point under consideration, and μ is
a pile–soil friction coefficient.

In undrained conditions (clay soils) and in term of total stress, $\mu = 0$ and one
obtains:

$$s = a = \alpha c_u \tag{4.4}$$

In other words, the unit shaft resistance is assumed equal to a fraction α of the
undrained strength of the soil. Some indications on the value of the coefficient α are
presented in Tables 4.2 and 4.3.

Table 4.2 Values of a, Eq. 4.4 (after Viggiani 1993)

Pile type	c_u (kPa)	a
Displacement	$c_u \leq 25$	1
	$25 \leq c_u \leq 70$	$1-0.0011(c_u-25)$
	$c_u \geq 70$	0.5
Replacement	$c_u \leq 25$	0.7
	$25 \leq c_u \leq 70$	$0.7-0.008(c_u-25)$
	$c_u \geq 70$	0.35

Table 4.3 Values of a, Eq. 4.4 (after Salgado 2008)

Pile type	a	Source
Displacement	$\left(\dfrac{c_u}{\sigma'_v}\right)_{NC}^{0.5}\left(\dfrac{c_u}{\sigma'_v}\right)^{-0.5}$ for $\dfrac{c_u}{\sigma'_v} \leq 1$	API (1993) Randolph and Murphy (1985) Salgado (2006)
	$\left(\dfrac{c_c}{\sigma'_v}\right)_{NC}^{0.5}\left(\dfrac{c_u}{\sigma'_v}\right)^{-0.25}$ for $\dfrac{c_u}{\sigma'_v} \geq 1$	
Replacement	0.55	O'Neill and Reese (1999)
	$0.4\left[1-0.12\ln\left(\dfrac{c_u}{p_A}\right)\right]^{0.55}$ for $3 \leq OCR \leq 5$	Hu and Randolph (2002); Salgado (2006)

Note
$p_A = 100\,\text{kPa} \cong 1\,\text{tsf} \cong 1\,\text{kgf/cm}^2$.

In drained conditions (sandy soils), and in terms of effective stress, $a=0$ and the unit shaft resistance may be expressed as:

$$s = \sigma'_h \mu = k\sigma'_v tg\delta \tag{4.5}$$

where $k = \dfrac{\sigma'_h}{\sigma'_v}$, σ'_v=effective overburden stress at the depth of the point under consideration, δ=friction angle at the pile–soil interface.

For displacement piles the horizontal stress existing before the installation of the pile increases as a consequence of the installation, and may reach failure in passive conditions; on the contrary, for replacement piles the horizontal stress decreases, and the active failure conditions may be approached.

It is to be remembered, however, that σ'_v (the overburden pressure existing before pile installation) is a principal stress, and is not equal to the vertical stress acting in the vicinity of the pile after installation, which is not a principal stress. The coefficient k, therefore, has to be determined empirically; it is dependent on the method of pile installation and the soil properties, though being necessarily in the range $k_a < k < k_p$.

Typical values of k and μ are suggested in Table 4.4. The choice of a design value leaves a wide margin to judgement. Poulos *et al.* (2001) list a number of additional suggestions, reported in Table 4.5.

Table 4.4 Suggested values of k and μ, Eq. 4.5 (after Viggiani 1993)

Pile type	Values of k for relative density		Values of μ
	Loose	*Dense*	
Displacement: steel H section	0.7	1.0	tg20° = 0.36
closed end pipe	1.0	2.0	
precast concrete	1.0	2.0	tg3φ/4
cast in place concrete	1.0	3.0	tgφ
Intermediate presso drill	0.7	0.9	tgφ
Replacement drilled shaft	0.5	0.4	tgφ
CFA	0.6	0.6	tgφ

Table 4.5 Suggestions for the evaluation of the coefficient k, Eq. 4.5

Pile type	Soil type	k	Reference
Replacement	Sand	$$k = \left[1 - \left(\frac{z}{L}\right)^{\alpha}\right] k_P + \left(\frac{z}{L}\right)^{\alpha} k_O$$ z = depth below surface; l = pile length; k_P = passive pressure coefficient; k_O = at rest pressure coefficient; $\alpha = 0.2$ (typically)	Yasufuku *et al.* (1997)
		$0.5 + 0.02\,N_{SPT}$	Go and Olsen (1993)
Displacement	Clay	$k = (1 - \sin\varphi')OCR^{0.5}$; φ' = effective angle of friction; OCR = overconsolidation ratio; also $\delta = \varphi'$	Burland (1973) Meyerhof (1976)
	Sand	$0.9 + 0.02\,N_{SPT}$	Go and Olsen (1993)

4.3.3 Large diameter bored piles

In the case of large diameter bored piles (drilled shafts), the ultimate base resistance is fully mobilized at displacements not less than 200 mm (see §4.2) and often much larger; the shaft resistance, on the contrary, is mobilized at much smaller displacements, around 20 mm. In dense or stiff soils the shaft resistance often exhibits unstable behaviour, in the sense that it develops a peak followed by a decrease towards a residual value. At displacements large enough to fully mobilize the base resistance, the shaft resistance can attain a value much smaller than the peak; accordingly, the ultimate pile capacity cannot be computed as the sum of the ultimate base and shaft capacities. Even if the shaft resistance increased monotonically up to the ultimate value and then kept constant, the displacements needed would be so high that a service load obtained by applying the usual safety factor would correspond usually to displacements that cannot be sustained by the structure.

To exemplify these considerations, let us refer to Figure 4.7, where schematic diagrams of the development of the resistances for different piles embedded in a homogeneous granular soil are reported.

For a medium diameter pile, either displacement or replacement (Figures 4.7a, 4.7b), the base resistance is usually a minor fraction of the ultimate pile resistance.

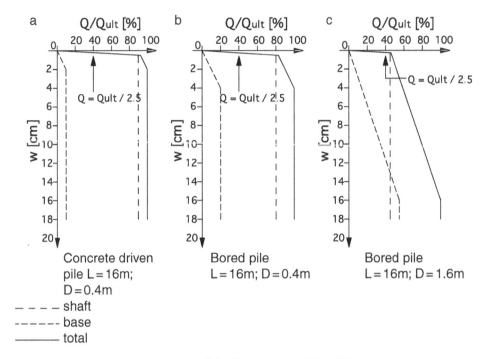

Figure 4.7 Schematic representation of the development of P and S.

At the service load, defined as Q_{lim}/FS, the fraction of the load taken by the base is very small, but the safety of the pile is not significantly affected.

For a large diameter bored pile (Figure 4.7c), the base resistance can be a major portion of the bearing capacity; in fact, it may be even larger than in Figure 4.7c, because the pile base is kept usually to better soils. It is evident that at service load the shaft resistance is fully mobilized and there is a small safety margin against very large displacements.

For these reasons the bearing capacity of large diameter bored piles is usually intended as a serviceability, rather than ultimate, limit state.

Berezantsev (1965) claims that the onset of plastic deformations around the pile base occurs at a displacement in the range $0.06d$ to $0.1d$, and suggested to assume the corresponding value of the base load in the evaluation of the bearing capacity. Such a load may be expressed as:

$$p = N_q^* \sigma_{vL}' \tag{4.6}$$

Eq. 4.6 is equivalent to Eq. 4.3, referring to the ultimate base load, but the values of N_q^* (Figure 4.8) are much smaller than the corresponding values of N_q.

Of course, shaft resistance may be evaluated by the same methods suggested for medium diameter bored piles (§4.3.2).

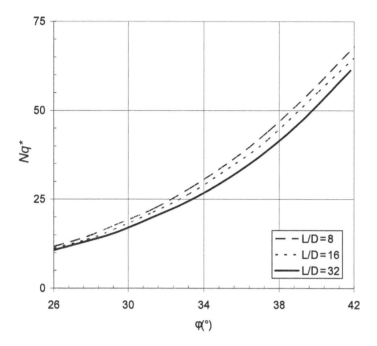

Figure 4.8 Values of the bearing capacity coefficient Nq^* corresponding to the onset of plastic deformations (source: Berezantzev 1965).

4.3.4 Micropiles

Due to their peculiarities, the bearing capacity of micropiles is still more dependent on the installation procedures than that of the other replacement piles.

Some authors (Salgado 2008) suggest evaluating the bearing capacity of micropiles in accordance with the suggestions given above for replacement piles of the CFA type. In the writers' experience, this criterion is overconservative because it neglects that micropiles are pressure injected. A merely empirical approach, such as that proposed by Bustamante and Doix (1985) may be preferable; it is widely adopted in France and other European countries.

First of all, a distinction is made between root piles (§2.2.5), that are called IGU (*injection globale unique*, i.e. single global injection) and the pile injected by a tube with valves, called IRS (*injection repetitive et selective*, i.e. repeated and selective injections). The soil characterization is made preferably by Ménard pressuremeter, but also by SPT; accordingly, the values of the lateral resistance s are given as a function of the limit pressure p_L of Ménard pressuremeter or SPT blowcount N_{SPT}.

It is further assumed that the shaft grouting pressure p_g is in the following ranges:

- $p_g \geq p_L$ for the IRS micropiles
- $0.5p_L \leq p_g \leq p_L$ for the IGU micropiles

and that the injection is carried out at a rate in the range 0.3 to 0.6 m³/h in cohesive soils, and 0.8 to 1.2 m³/h in cohesionless soils.

The shaft resistance of a micropile is expressed as:

$$S = \pi d_s L_s s$$

where d_s is the expanded diameter, L_s the length of the injected portion of the pile and s is the shear resistance at the interface between the soil and the injected portion of the pile (Figure 4.9). The expanded diameter is expressed as $d_s = \alpha d$; the values of the factor α are given in Table 4.6.

The shear resistance s at the interface between the soil and the injected portion of the pile is given as a function of p_L or N_{SPT} by the following expressions:

$$s = a + b p_L \tag{4.7}$$

$$s = \alpha + \beta N_{SPT} \tag{4.8}$$

where the values of the parameters a, b, α and β are given in Table 4.7.

If the injected portion of the micropile extends until the soil surface, it is recommended that the upper 5 m are considered in any case as if they were of the IGU type. It is also recommended that the injected length L_s, not considering the upper 5 m, should be not less than 4 m; this means that the total length of a micropile of the IRS type should be at least 9 m.

The point resistance is usually assumed equal to 15% of the shaft resistance; therefore:

$$Q_{ult} = 1.15 s L_s \pi \alpha d$$

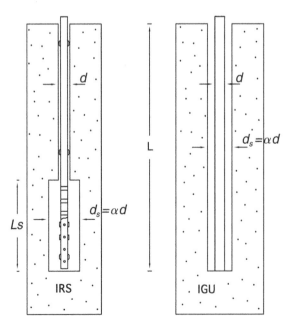

Figure 4.9 Scheme of an injected micropile.

Table 4.6 Values of the coefficient α to obtain the expanded value of the diameter of a micropile

Type of soil	Values of α		Minimum suggested grout volume*
	IRS	IGU	
Gravel	1.8	1.3–1.4	$1.5\,V_S$
Sandy gravel	1.6–1.8	1.2–1.4	$1.5\,V_S$
Gravelly sand	1.5–1.6	1.2–1.3	$1.5\,V_S$
Coarse sand	1.4–1.5	1.1–1.2	$1.5\,V_S$
Medium sand	1.4–1.5	1.1–1.2	$1.5\,V_S$
Fine sand	1.4–1.5	1.1–1.2	$1.5\,V_S$
Silty sand	1.4–1.5	1.1–1.2	IRS: $(1.5–2)V_S$; IGU: $1.5\,V_S$
Silt	1.4–1.6	1.1–1.2	IRS: $2\,V_S$; IGU: $1.5\,V_S$
Clay	1.8–2.0	1.2	IRS: $(2.5–3)V_S$; IGU: $(1.5–2)V_S$
Marl	1.8	1.1–1.2	$(1.5–2)V_S$ for stiff layers
Marly limestone Weathered or fractured limestone	1.8	1.1–1.2	$(2–6)V_S$ or more for fractured layers
Weathered and/or fractured rock	1.2	1.1	$(1.1–1.5)V_S$ for slightly fractured layers $2\,V_S$ or more for fractured layers

Note

$$V_S = L_S \frac{\pi d_S^2}{4}$$

Table 4.7 Values of the coefficients in Eqs. 4.7 and 4.8

Soil type	Micropile type	a (MPa)	b	α (MPa)	β (MPa)
Sand and gravel	IGU	0	0.10	0	0.005
	IRS	0.05	0.10	0.05	0.005
Silt and clay	IGU	0.04	0.06	0.04	0.004
	IRS	0.10	0.084	0.1	0.006
Weathered and fractured rock	IGU	0.04	0.10	–	–
	IRS	0.04	0.13	–	–

4.4 Bearing capacity from correlation with penetrometer data

4.4.1 CPT

It is assumed that:

$$p = c_p q_{cL}; \quad s_i = c_{si} q_{ci} \tag{4.9}$$

where c_p and c_{si} are empirical coefficients that depend on soil type and pile type; q_{cL} is a value of the point resistance of CPT representative of the pile base level; q_{ci} is the cone resistance representative of the layer i. The value of q_{cL} is obtained by averaging q_c in a suitable interval of depth around the pile base: for instance, between $(L–1.5d)$ and $(L+1.5d)$, provided there are no strong stratigraphic changes within a range of five to eight pile diameters above the pile base. Particular care should be taken where piles are driven through weak materials to penetrate only one or two diameters into a dense layer (Meyerhof and Valsangkar 1977). The value of q_{ci} is the average of q_c over the layer i.

It has been often assumed that, for driven piles in sand, $c_p = 1$ and hence $p = q_{cL}$ (Fleming *et al.* 1985; Viggiani 1993). Recent experimental evidence and analyses (Jardine and Chow 1996; Randolph 2003; White 2003) seems to indicate that smaller values are more proper. Typical values suggested for the coefficient c_p are listed in Table 4.8 for piles in sand and Table 4.9 for piles in clay.

Some indications on the values of the coefficient c_s are reported in Table 4.10 for piles in sand, and Table 4.11 for piles in clay.

Table 4.8 Pile base resistance p vs. CPT cone resistance q_c for piles in sand (Eq. 4.9)

Pile type	c_p	Notes	Source
Displacement	0.35÷0.5	Database of high quality pile load tests	Chow (1997)
	0.20÷0.35	Computed	Lee and Salgado (1999)
	0.32÷0.47	Test data	
	0.4	Reinterpretation of the Chow (1997) data	Randolph (2003)
	0.4 for Franki piles 0.57 for precast concrete piles	Data from pile load test; Q_{lim} by Van der Veen's criterion	Aoki and Velloso (1975)
Replacement	0.2	Load tests on drilled shafts	Franke (1989)
	0.13±0.02	Calibration chamber load tests	Ghionna *et al.* (1994)
	0.23 exp(−0.0066D$_R$)	FEM analyses and calibration chamber tests	Salgado (2006)
	0.20÷0.26	Test data	Lee and Salgado (1999)

Table 4.9 Pile base resistance p vs. CPT cone resistance q_c for piles in clay (Eq. 4.9)

Pile type	c_p	Notes	Source
Displacement	0.9÷1.0	Soft to lightly OC clays	State of the art
	0.35 for driven piles 0.30 for jacked piles	Stiff clays	Price and Wardle (1982)
Replacement	0.47 for pure clay 0.52 for silty clay 0.78 for silty clay with sand 0.71 for sandy clay with silt 0.83 for sandy clay	Medium to stiff clays	Aoki and Velloso (1975) Aoki *et al.* (1978)
	0.34 for pure clay and silty clay 0.41 for silty clay with sand and sandy clay with silt 0.66 for sandy clay	Medium to stiff clays	Lopes and Laprovitera (1988)

Table 4.10 Pile shaft resistance s vs. CPT cone resistance q_c for piles in sand (Eq. 4.9)

c_s	Source
0.008 for open ended steel pipe piles 0.012 for precast concrete and closed-ended steel pipe piles	Schmertmann (1978)
0.004 ÷ 0.006 per D_R ≤50% 0.004 ÷ 0.007 per 50% < D_R ≤70% 0.004 ÷ 0.009 per 70% < D_R ≤90% Closed-ended pipe piles	Lee *et al.* (2003)
0.0040 for clean sand 0.0057 for silty sand 0.0069 for silty sand with clay 0.0080 for clayey sand with silt 0.0086 for clayey sand Driven piles: for Franki piles: multiply number above by 0.7 For drilled shafts: multiply number above by 0.5	Aoki and Velloso (1975) Aoki *et al.* (1978)
0.0027 for clean sand 0.0037 for silty sand 0.0046 for silty sand with clay 0.0054 for clayey sand with silt 0.0058 for clayey sand Replacement piles	Lopes and Laprovitera (1988)
0.0034 ÷ 0.006 This method uses a corrected value of cone resistance q_c–u, where u is the pore pressure at the depth considered	Eslami and Fellenius (1997)

Table 4.11 Pile shaft resistance s vs. CPT cone resistance q_c for piles in clay (Eq. 4.9)

c_s	Source
0.074 ÷ 0.086 for sensitive clay 0.046 ÷ 0.056 for soft clay 0.021 ÷ 0.028 for silty clay or stiff clay Driven piles This method uses a corrected value of cone resistance q_c–u, where u is the pore pressure at the depth considered	Eslami and Fellenius (1997)
0.025 Displacement piles	Thorburn and MacVicar (1971)
0.017 for pure clay 0.011 for silty clay 0.0086 for silty clay with sand 0.0080 for sandy clay with silt 0.0069 for sandy clay Driven piles: for Franki piles: multiply number above by 0.7 For drilled shafts: multiply number above by 0.5	Aoki and Velloso (1975) Aoki *et al.* (1978)
0.012 for pure clay 0.011 for silty clay 0.010 for silty clay with sand 0.0087 for sandy clay with silt 0.0077 for sandy clay Non displacement piles	Lopes and Laprovitera (1988)

4.4.2 SPT

It is assumed that:

$$\frac{p}{p_A} = n_p N_L; \quad \frac{s}{p_A} = n_{si} N_i \tag{4.10}$$

where n_p and n_{si} are empirical coefficients that depend on soil type and pile type; N_L is a value of the blow count SPT representative of the pile base level; N_i is the blow count representative of the layer i; p_A is a reference pressure $= 100\,\text{kPa} \cong 1\,\text{tsf} \cong 1\,\text{kgf/cm}^2$. The value of N_L is obtained by averaging N in a suitable interval of depth around the pile base: for instance, between $(L-1.5d)$ and $(L+1.5d)$. Of course, since values of N are obtained at discrete depths, only few values of N will be usually available for averaging in the calculation of N_L so that some judgement is needed in the averaging process.

Typical values suggested for the coefficient n_p are listed in Table 4.12 for piles in sand and Table 4.13 for piles in clay. Tables 4.14 and 4.15 report the values of n_s respectively for piles in sand and in clay.

4.5 Driving formulas

The bearing capacity of a driven pile may be related to the measured permanent displacement (or "set") δ of the pile under a hammer blow and to the energy of the hammer, through the so-called "driving formulas". These are based on a balance

Table 4.12 Suggested values of n_p for piles in sand (Eq. 4.10)

Pile type	n_p	Source
Displacement	4 4.8 for clean sand 3.8 for silty sand 3.3 for silty sand with clay 2.4 for clayey sand with silt 2.9 for clayey sand For Franki piles: multiply numbers above by 0.7	Meyerhof (1983) Aoki and Velloso (1975)
	3.25 for sand 2.05 for sandy silt 1.65 for clayey silt 1.00 for clay	Decourt (1995)
Replacement	0.82 for clean sand 0.72 for sand with silt or clay	Lopes and Laprovitera (1988)
	0.6 ($p/p_A \leq 45$)	Reese and O'Neill (1989)
	1.9 for CFA piles 1.2 < for drilled shafts	Neely (1991)
	1.65 for sand 1.15 for sandy silt 1.00 for clayey silt 0.080 for clay	Decourt (1995)

Table 4.13 Suggested values of n_P for piles in clay (Eq. 4.10)

Pile type	n_P	Source
Displacement	0.95 for pure clay	Aoki and Velloso (1975)
	1.05 for silty clay	Aoki *et al.* (1978)
	1.57 for silty clay with sand	
	1.43 for sandy clay with silt	
	1.67 for sandy clay	
	For Franki piles: multiply numbers above by 0.7	
Replacement	0.47 for pure clay	Aoki and Velloso (1975)
	0.52 for silty clay	Aoki *et al.* (1978)
	0.78 for silty clay with sand	
	0.71 for sandy clay with silt	
	0.83 for sandy clay	
	0.34 for pure clay and silty clay	Lopes and Laprovitera
	0.41 for silty clay with sand and sandy clay with silt	(1988)
	0.66 for sandy clay	

Table 4.14 Suggested values of n_S for piles in sand (Eq. 4.10)

Pile type	n_S	Source
Displacement	0.02 ($s \leq 100\,kPa$)	Meyerhof (1976) Thorburn and MacVicar (1971)
	0.02 for full displacement piles	Meyerhof (1976, 1983)
	0.01 for H piles	
	0.033 for sand	Aoki and Velloso (1975)
	0.038 for silty sand	Aoki *et al.* (1978)
	0.040 for silty sand with clay	
	0.033 for clayey sand with silt	
	0.043 for clayey sand	
	For Franki piles: multiply numbers above by 0.7	
Replacement	0.01 ($s \leq 50\,kPa$)	Meyerhof (1976)
	0.016 for sand	Aoki and Velloso (1975)
	0.019 for silty sand	Aoki *et al.* (1978)
	0.020 for silty sand with clay	
	0.016 fror clayey sand with silt	
	0.021 for clayey sand	
	0.014 for sand	Lopes, Laprovitera (1988)
	0.016 for silty sand	
	0.020 for silty sand with clay	
	0.024 fror clayey sand with silt	
	0.026 for clayey sand	

between the input energy of the hammer L_i, the work L_u required to move the pile permanently at a distance δ and the work L_d dissipated in the impact.

The input energy $L_i = \eta E$, where η is an efficiency and E is the theoretical energy of a hammer blow. For a drop hammer of weight W and height of fall h, $E = Wh$; for steam or diesel hammer, it is a quantity given by the manufacturer.

Table 4.15 Suggested values of n_S for piles in clay (Eq. 4.10)

Pile type	n_S	Source
Displacement	0.029 for clay 0.021 for silty clay 0.024 for silty clay with sand 0.020 for sandy clay with silt and sandy clay For Franki piles: multiply number above by 0.7	Aoki and Velloso (1975) Aoki *et al.* (1978)
Replacement	0.014 for clay	Aoki and Villoso (1975)
	0.010 for silty clay 0.012 for silty clay with sand 0.010 for sandy clay with silt and sandy clay	Aoki *et al.* (1978)
	0.024 for clay 0.022 for silty clay 0.024 for silty clay with sand 0.022 for sandy clay with silt 0.031 for sandy clay Non-displacement piles	Lopes and Laprovitera (1988)

Assuming that the pile moves as a rigid body and that the resistance of the pile to the dynamic action of the hammer is equal to the static bearing capacity Q, the work L_u may be expressed as (Figure 4.10):

$$L_u = Q_{ult}\delta$$

If the dissipated work is assumed to be nil, one obtains $Q_{ult}\delta = \eta E$, and hence:

$$Q_{ult} = \frac{\eta E}{\delta}$$

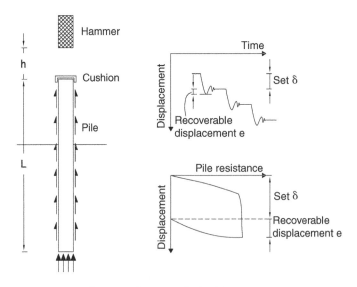

Figure 4.10 Schematic diagram of pile driving.

which is known as Sanders formula, and dates back to 1857. Other formulas are obtained by different evaluations of the dissipated work. For instance, taking into account the work dissipated in elastic rebound e, equal to $Q_{ult}\, e/2$, one obtains:

$$Q_{ult} = \frac{\eta E}{\delta + \dfrac{e}{2}}$$

which is widely used with $\eta = 1$ and $e = 2.5\,\text{mm}$ as the Engineering News formula (Wellington 1893).

Taking into account the energy loss due to the deformation of the pile, the cushion and the soil, and using the Newton theory of the impact between two bodies, the balance may be written:

$$\eta E = Q_{ult}\delta + \eta E(1 - n^2)\frac{W_p}{W + W_p} + \frac{Q_{ult}}{2}(C_1 + C_2 + C_3) \tag{4.11}$$

where: n is the impact restitution coefficient ($n = 1$ for a perfectly elastic impact, $n = 0$ for a totally inelastic one); W_p is the weight of the pile and the three terms C_1, C_2 and C_3, multiplied by $Q/2$, represent the work dissipated in the compression of the pile, cushion and soil. From Eq. 4.11 one obtains:

$$Q_{ult} = \frac{\eta E}{\delta + \dfrac{C_1 + C_2 + C_3}{2}} \frac{W + n^2 W_p}{W + W_p} \tag{4.12}$$

which is known as the Hiley (1925) formula. Generally it is assumed that $C_3 = 2.5\,\text{mm}$; C_1 may be evaluated as the elastic compression of the pile under the axial load Q; the value of η, n and C_2 are listed in Tables 4.16, 4.17 and 4.18 respectively.

Table 4.16 Values of η in the pile driving formulas

Type of hammer	Values of η
Drop hammer, triggered fall	1
Steam or compressed air	0.9
Drop hammer, winch operated	0.8
Diesel	0.7–0.8

Table 4.17 Values of the coefficient of restitution n

Conditions	Values of n
Timber pile, damaged top	0
Timber pile, intact top	0.3
Greenheart oak	0.5
Precast concrete, no cushion	0.4
Micarta plastic	0.8

Table 4.18 Values of C_2, Eq. 4.12

Pile type	Values of C_2 (mm)			
$4Q/\pi d^2$ (MPa) 3.5		7.0	10.5	14.0
Steel, no cushion	0	0	0	0
Wood, no cushion	1	2.5	4	5
Precast concrete with cushion	3	6	9	15
Steel with wood cushion	1	2	3	4

In clayey soils the evaluation of the bearing capacity by driving formulas should be based on the set obtained by redriving the piles when the excess pore pressure generated by the first driving has dissipated. This requires a time interval in the range between a few days and a couple of weeks.

Finally, it should be emphasized that driving formulas are rather uncertain and unreliable; available comparisons between their results and the actual bearing capacity, as determined by load tests (Sørensen and Hansen 1956; Agerschou 1962), show a wide scatter, such to raise serious doubts on their use.

4.6 The wave equation analysis

A better description of the phenomena occurring in the pile during driving may be obtained by modelling the pile as an elastic rod subjected to the impact of the hammer, which generates a stress wave travelling down the pile. Any resistance to the movement of the pile, for example due to shaft friction or defects in the pile, causes an upward travelling wave to be propagated back up towards the pile head; the same occurs at the pile tip.

The stress wave equation is:

$$\frac{\partial^2 w}{\partial t^2} = c^2 \frac{\partial^2 w}{\partial z^2} + f \tag{4.13}$$

where $c = \sqrt{\dfrac{E}{\rho}}$ is the velocity of the longitudinal stress wave propagation in the bar; E and ρ are the Young modulus and the density of the pile material; $f = \pi ds$ is the mobilized soil resistance per unit length of pile.

In the analysis of Eq. 4.13 the pile–soil interface is usually modelled as an elastic, perfectly plastic spring in parallel with a linear dashpot (Smith 1960). The latter simulates not only the viscous damping but also the radiation damping (mass of the soil and associated inertia effects which occur in reality and are neglected in the model).

The equation is solved numerically by a lumped mass model.

Acceleration and strain may be recorded in the vicinity of the pile top during driving; generally pairs of instruments (accelerometers and strain gauges) are fixed to the pile in diametrically opposite positions, in a section at least two diameters away from the pile top, and the readings averaged to eliminate possible bending effects. The strain data are converted to force by multiplying the measured strain by the axial rigidity EA of the pile, while the acceleration data are converted in velocity and displacement by integration.

As mentioned above, soil resistance along the shaft and at the toe of the pile causes reflected waves that travel back and are felt at the pile top. Of course, the time at which the reflected waves arrive at the pile top depends on the location along the shaft of the soil reactions generating them. In principle, the force and velocity measured at the pile top thus provide information to estimate soil resistance and its distribution. For a given measured variation of the pile head velocity with time, the soil parameters are adjusted until the measured force variation is simulated accurately by the computer model. Generally there are a number of different distributions of soil parameters giving a satisfactory fit to the measured response, so that a good deal of subjective judgement is required.

Since viscous and inertia effects are lumped into a single parameter, the total soil resistance evaluated by the stress wave analysis in the above hypothesis includes static and viscous components. Further hypotheses are needed to separate the two components and estimate the static resistance Q_{ult}.

The Case Western Reserve University, Ohio, has developed instruments and software for monitoring the driving operations and for high strain dynamic testing (CASE method). Instruments include pairs of accelerometers and strain transducers to be fixed to the pile head; the data are recorded by a pile driving analyser (PDA) which is basically a signal conditioning and amplification device, carrying out also the integrations and multiplications. The output is in terms of peak acceleration, velocity, force and energy, and the calculated static pile resistance. A number of data have to be input before analysis (hammer features, pile length and material, stiffness and damping parameters of the soil), based on previous experience and knowledge of the nature of the soil. This make the evaluation of static capacity rather uncertain, unless correlation with a static load test at the site is available.

The data recorded by PDA may be analysed by the CAPWAP software, based on the wave equation; the analysis predicts the load–settlement curve and the bearing capacity of the pile.

Randolph (2003) believes that the use of wave equation to predict the static behaviour of the pile (from driving data or high energy testing) is promising, provided better models of the dynamic pile – soil interaction is developed and implemented into proper software.

4.7 Bearing capacity of pile groups

The bearing capacity of a pile group, Q_{Gult}, is generally expressed as the bearing capacity Q_{ult} of the single pile multiplied by the number n of piles in the group and by a coefficient E called "efficiency" of the group:

$$Q_{Gult} = EnQ_{ult} \tag{4.14}$$

The proper value of the efficiency is selected on the basis of the available experimental evidence, essentially gathered by small-scale natural gravity or centrifuge tests.

With the usual values of the spacing axis to axis s among piles ($2.5 \leq s/d \leq 3.5$), in cohesionless soils and for displacement piles the efficiency is always greater than unity, and approaches unity at a spacing exceeding $6d$. Even for replacement piles, in cohesionless soils the efficiency is generally not less than unity. Accordingly, in the design evaluation of the bearing capacity of a pile group in cohesionless soils, the efficiency can be conservatively assumed equal to unity (no group effect).

In cohesive soils and in undrained conditions ($\varphi = 0$ analysis in terms of total stress), the efficiency is always less than unity, and again approaches unity at a spacing around six times the diameter. It may be evaluated either by empirical expressions as a function of the group geometry, or considering the pile group and the soil in between as a block foundation.

One of the empirical expressions is the Converse-Labarre formula:

$$E = 1 - \frac{arctg \dfrac{s}{d}[(f-1)g + (g-1)f]}{\dfrac{\pi}{2}fg}$$

where f and g represent the number of rows and columns of piles in the group ($fg = n$). Another popular expression is the so-called Feld's rule, which states that the efficiency of each pile in a group is reduced by 1/16 for each adjacent pile along row, column or diagonal. The efficiency of the group is the average of that of all the piles. Figure 4.11 clarifies the rule.

Terzaghi and Peck (1948) suggest evaluating the bearing capacity of the group by block failure (Figure 4.12) with the expression:

$$Q_{G\lim} = B_1 B_2 (N_c c_u + \gamma L) + 2l(B_1 + B_2)c_u \tag{4.15}$$

where $B_1 = fs$ and $B_2 = gs$ are the overall width and length of the group ($B_1 \leq B_2$). The bearing capacity coefficient N_c may be given the values reported in Table 4.19.

If the pile group is required to carry the full working load within a few days or weeks after the piles have been installed, the remoulded undrained shearing strength should be taken for c_u in the second term of Eq. 4.15, representing the lateral resistance, while in the first term the undisturbed strength can be used, since the greater part of the soil below the group remains in an undisturbed state.

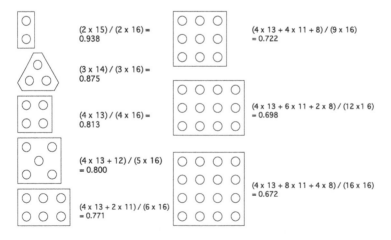

Figure 4.11 Feld's rule.

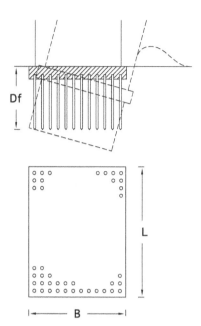

Figure 4.12 Block failure of a pile group.

Table 4.19 Values of N_c in Eq. 4.15

L/B_1	N_c	
	$B_2/B_1=1$	$B_2/B_1 \geq 10$
0.25	6.7	5.6
0.50	7.1	5.9
0.75	7.4	6.2
1.00	7.7	6.4
1.50	8.1	6.8
2.00	8.4	7.0
2.50	8.6	7.2
3.00	8.8	7.4
≥ 4	9.0	7.5

4.8 Rock socketed piles

4.8.1 Introduction

Rock socketed bored piles may be a cost-effective solution for foundations which carry large column loads, when bedrock covered by weak soils is found within a reasonable depth.

Rock sockets have a number of features which differentiate them from other types of piles. The generally stubby geometry should imply a more even distribution of capacity between shaft and base, with the latter carrying a significant fraction of the applied load. However, the low ratio of pile modulus to rock modulus leads to high relative compressibility and this factor, coupled with a tendency for strain-softening behaviour of the load transfer response along the shaft, produces an overall response where the shaft capacity may be fully mobilized, and potentially degraded, before any significant mobilization of the base load. Such behaviour is further accentuated by the usual occurrence of a thin layer of compressible sediments at the bottom of the hole, as a consequence of an imperfect cleaning before concreting, and/or of a frequent zone of weaker concrete just over the pile base.

Figure 4.13 reports the load sharing between the shaft and the base of the socket, evaluated by modelling the rock as an elastic half space and the pile as an elastic cylinder with a final softer zone to simulate either the presence of sediments at bottom or defect in the pile due to the presence of weaker concrete. It is evident that, for a socket length $l \geq 4d$, practically no load is transferred to the pile base in competent intact rock ($E_C/E_R \leq 5$ where E_C is the concrete Young's modulus and E_R is the rock modulus) and even for a very weak rock the fraction of load carried by the base is only about 15%. For $l/d = 2$, these fractions increase respectively to 12% and 26%, but they decrease dramatically when a defect occurs near the pile base. Such findings are supported by a broad experimental evidence (Osterberg and Gill 1973; Chang and Wong 1987; Radhakrishnan and Leung 1989; Kulhawy and Phoon 1993; Leung and Randolph 2005).

For these reasons it is suggested that, in general, the base resistance is neglected in the evaluation of the bearing capacity of a rock socketed pile. It can be taken into account only for dry boreholes of large diameter and at shallow depth, where effective inspection and cleaning is possible.

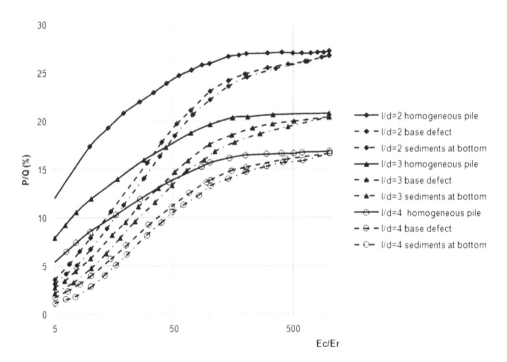

Figure 4.13 Load sharing between shaft and base for a socketed pile (elastic analysis).

4.8.2 *Shaft resistance*

Seidel and Cho (2002) point out that the peak shear resistance of rock sockets is the result of a complex interaction of the following parameters:

- the intact shear strength of the rock;
- the residual friction angle;
- the mass modulus and Poisson's ratio of the surrounding rock mass;
- the hydrostatic pressure of the wet concrete;
- the diameter of the socket; and
- the roughness of the socket wall.

To predict the shaft resistance of a rock socket, a constitutive model for interface sliding is hence required, that incorporates the coupling of shear and normal modes of displacement (Pease and Kulhawy 1984; Seidel and Haberfield 1995) and permits the description of the development of shaft resistance from initial loading until full slip. These models, however, require accurate numerical values of parameters which are not available in routine engineering practice; accordingly, it is customary to refer to methods correlating the ultimate shaft friction s to some rock properties like the uniaxial compressive strength q_u (Horvath *et al.* 1980; Williams and Pells 1981; Amir 1986; Rowe and Armitage 1987; Kulhawy and Phoon 1993).

Kulhawy and Phoon (1993) collected a number of data in terms of adhesion factor α ($\alpha = s/c_u$ for piles in clay, or $\alpha = s/0.5q_u$ for piles in rock), and proposed the following empirical relationships (Figure 4.14):

$$\alpha = \Psi \left[\frac{c_u}{p_a} \right]^{-0.5} \tag{4.16a}$$

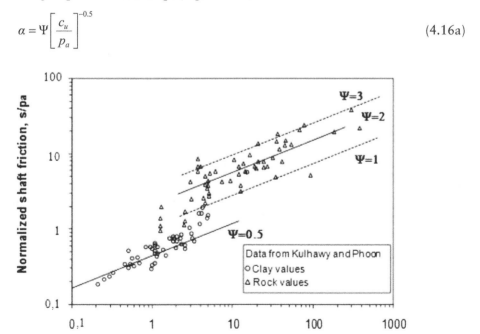

Figure 4.14 Relation between socket reduction factor α and normalised rock strength (after Kulhawy and Phoon 1993).

and hence:

$$s = \Psi[c_u p_a]^{0.5} \tag{4.16b}$$

for clays, and

$$\alpha = \Psi\left[\frac{q_u}{2p_a}\right]^{-0.5} \tag{4.16c}$$

and hence:

$$s = \Psi\left[\frac{q_u}{2}p_a\right]^{0.5} \tag{4.16d}$$

for rocks.

The values of Ψ are listed in Table 4.20.

It is important to emphasize that the data reported in Figure 4.14 are site-averaged; understandably, the results of individual load tests showed greater scatter than the site-averaged data. The coefficient of correlation of the empirical relationships given above is 0.71 for the averaged data but only 0.46 for the individual data.

4.8.3 Base resistance

An extensive database presented by Zhang and Einstein (1998) shows that the end-bearing capacity mobilized at a displacement of 10% of the pile diameter can be assumed to be proportional to the square root of the uniaxial compressive strength:

$$(q_{ult} / p_a) \sim 15(q_u / p_a)^{0.5}$$

in which p_a is the atmospheric pressure. This expression fits some experimental results of model piles in calcarenite by Randolph (1998), as shown in Figure 4.15.

If a more careful evaluation of the base resistance is believed to be worthwhile, it may be performed following the indication below.

The typical bearing capacity failure modes are shown in Figure 4.16 and depend on the discontinuity spacing or layering.

For a thick rigid layer overlying a weaker one, failure may be by flexure. The flexural strength is of the order of 10–20% of the compressive strength.

For a thin layer overlying a weaker one, failure can be by punching that, in effect, is manifested by a tensile failure in the rock material. It is important to realize that,

Table 4.20 Values of Ψ in Eq. 4.16

Type of material		Ψ
Clay		0.5
Rock	Lower bound	1
	Mean	2
	Upper bound	3

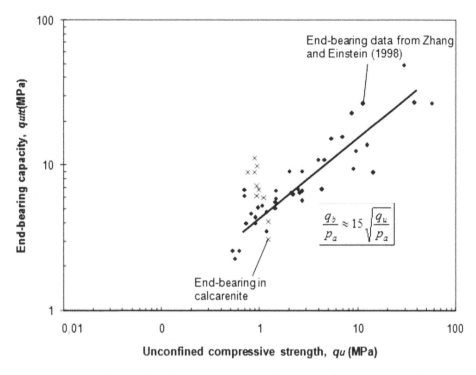

Figure 4.15 Correlation of end bearing capacity of piles in rock and unconfined compression strength of the rock.

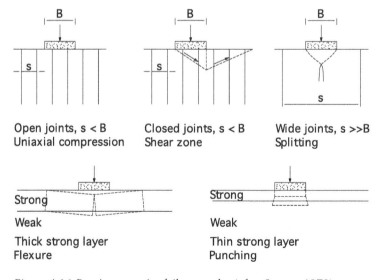

Figure 4.16 Bearing capacity failure modes (after Sowers 1979).

in both of these cases, failure of the underlying layer could occur first by one of the other failure modes.

For loading applied to a rock mass with open joints, where the joint spacing S is less than the pile diameter D, failure is likely to occur by uniaxial compression of rock columns. If the rock mass behaviour is idealized as a cohesive-frictional material, the ultimate bearing capacity q_{ult} is given by:

$$q_{ult} = q_u = 2c\tan(45° + \varphi/2)$$

in which q_u = uniaxial compressive strength, c and φ are the cohesion and the friction angle respectively of the rock mass (rather than intact rock) properties.

If the rock mass contains closely spaced closed joints a general wedge type mode of failure may develop.

The ultimate bearing capacity q_{ult} in this case is given by the well-known solution for a shallow foundation, adapted for a cylinder with diameter D at depth L:

$$q_{ult} = (1 + \tan\varphi)\gamma LN_q + (1 + N_q/N_c)cN_c + 0.6D\gamma N_\gamma/2$$

in which γ is the unit weight of the rock and N_q, N_c and N_γ are bearing capacity factors given in Figure 4.17.

The second term is normally small compared to the other terms and is often neglected.

For cases in which the joints are spaced more widely than the pile diameter D, failure may occur by splitting beneath the tip, which eventually leads to general

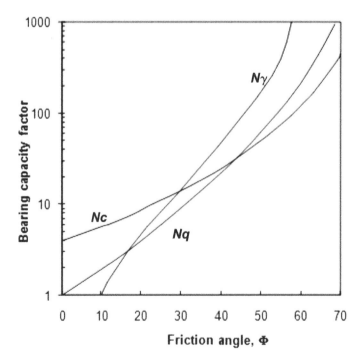

Figure 4.17 Wedge bearing capacity factors.

shear failure. This problem has been evaluated by Kulhawy and Goodman (1980) assuming no stress is transmitted across the vertical discontinuity, to give:

$$q_{ult} \sim JcN_{cr}$$

in which N_{cr} is a bearing capacity factor given in Figure 4.18 and J is a correction factor given in Figure 4.19.

In the evaluation of q_{ult} as outlined above, it is important to note that the strength parameters of the rock mass, and not of the intact rock, must be used. Values of c

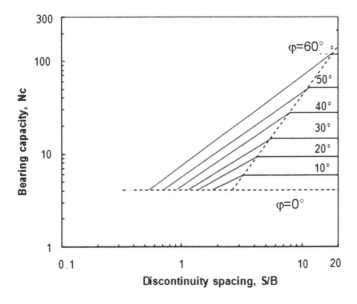

Figure 4.18 Bearing capacity factors for open joints (after Kulhawy and Goodman 1980).

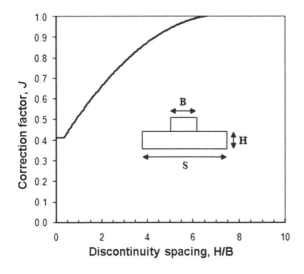

Figure 4.19 Correction factor for discontinuity spacing (after Bishnoi 1968).

and φ obtained for the intact rock material are considerably higher than those for rock mass, and their use will result in an overestimate of the actual bearing capacity.

If the actual rock mass properties are not evaluated, Kulhawy and Goodman (1987) suggested that values for c or q_u for the intact rock may be reduced as shown in Table 4.21.

Table 4.21 Reduction of strength parameters for rock mass

RQD [%]	Rock mass properties		
	Uniaxial compressive strength	Cohesion, c	Friction angle, φ
0–70	$0.33\,q_u$	$0.10\,q_u$	30°
70–100	$(0.33–0.80)\,q_u$	$0.10\,q_u$	30°–60°

5 Settlement

5.1 Introduction

As mentioned in the introduction, the present day routine design practice for pile foundations neglects the contribution of the cap, assuming that the whole load is transmitted to the soil through the piles. Most codes and regulations suggest or prescribe this approach. Due to this conservative assumption and to the circumstance that piles generally cross relatively deformable shallow soils to reach deep less deformable layers, settlement of pile foundations is generally small and is seldom considered a significant design issue.

Some experimental results referring to settlement of full scale foundations and small scale physical models in London Clay (Cooke 1986) throw further light on this topic. Figure 5.1a shows the settlement at the end of primary consolidation of full-scale unpiled rafts; to account for the influence of foundation shape, the settlement is plotted against the breadth B of an equivalent square foundation, obtained

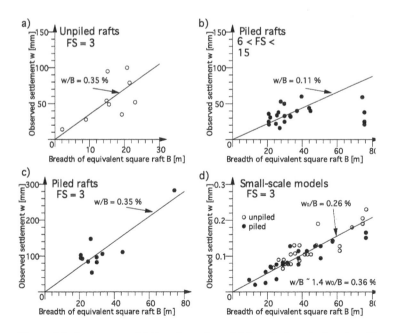

Figure 5.1 Settlement of full scale and model foundations in London Clay.

by multiplying the actual breadth by the ratio of the settlement influence factor for the appropriate rectangular raft to that for a square raft. The best straight line through the points corresponds to a settlement of 0.35%B. It may be surprising that the load q does not enter the calculation; this finding may be explained by assuming that all the foundations considered have nearly the same safety factor FS. The service load may be then expressed as $q = N_c c_u / FS$, where N_c is a bearing capacity factor and c_u is the undrained strength of London Clay. As it is well known, the undrained modulus E_u of a clay may be assumed roughly proportional to its undrained strength by the relation $E_u = k c_u$, with the ratio k depending on the OCR and nature of the clay (Table 5.1), and hence constant for London Clay. The immediate settlement w_o of a foundation of breadth B may be then expressed as:

$$w_o = \frac{qB}{E_u} I_w = \frac{N_c I_w}{kFS} B$$

Being the final settlement proportional to the immediate one, it may be concluded that the observed settlement of rafts on London Clay are related to the raft width B in a manner that is consistent with the elastic theory.

Figure 5.1b shows the observed settlement of structures on piled foundations plotted once more against the breadth of the equivalent square foundation. This time the best straight line through the point is $w = 0.11\%B$; this seems to suggest that the settlements of structures on deep foundations are of the order of one-third of those of comparable structures on shallow foundations. All these foundations had been designed with a nominal safety factor of the order of three but, due to the conservative assumptions adopted for piled foundations, the actual safety factors were computed by Cooke in the range 6 to 15. Accordingly, Cooke suggests correcting the observed settlement to obtain the values corresponding to a safety factor of three. Referring to the original paper for the details of the correction procedure, Figure 5.1c reports the results obtained; the relationship between the settlement and the breadth is identical with that derived for structures on unpiled rafts.

As a further confirmation, in Figure 5.1d the immediate or undrained settlement of small-scale model foundations with a safety factor of three are reported; it is equal to 0.26% of the foundation size B, irrespective of the foundation being a piled or unpiled one. Being the final settlement of the order of 1.4 times the immediate one, model tests led to results similar to full-scale observations.

From these and much other evidences it could be concluded that the settlement of a pile foundation, designed according to the presently most usual approach, could safely be neglected and the methods for its prediction have only an academic interest. Such a conclusion, however, has to be corrected for at least three reasons.

Table 5.1 Values of the ratio E_u/c_u

OCR	E_u/c_u		
	$I_p < 0.3$	$0.3 < I_p < 0.5$	$I_p > 0.5$
< 3	800	400	200
3–5	500	300	150
> 5	300	200	100

1   Settlement of a piled foundation may be significant for very large foundations and/or when there are deep layers of compressible soils; the well-known case of Charity Hospital in New Orleans quoted by Terzaghi and Peck (1948) is an instructive example (Figure 5.2).

2   Most codes and regulations require the settlement be kept below some admissible value; to prove that this is actually the case, settlement predictions have to be carried out.

3   Some of the methods of analysis that are employed in the prediction of settlement are fundamentally the same to be adopted for the analysis of soil – structure interaction and the structural design of the cap; furthermore, they are the basis for the design methods of piled rafts that will be presented in Part IV of this book.

Most of the methods for the prediction of the settlement of pile foundations have as a first step the prediction of the settlement of a single pile; it is therefore convenient to start from this topic.

5.2 Settlement of the single pile

5.2.1 Empirical methods

The most reliable means to study the load settlement behaviour of a foundation pile is that of performing a full-scale load test on a prototype pile (Chapter 7). Such a test, unfortunately, cannot be run at the design stage but in particular cases, because of its high cost in terms of resources, time and different constraints.

A typical load settlement relationship, as determined in a load test, has been reported in Figure 4.2. It is characterized by an evident non-linearity, even in the initial stages of loading. Such a feature is common to practically any pile type in any soil, with the only possible exception of piles point bearing on rock.

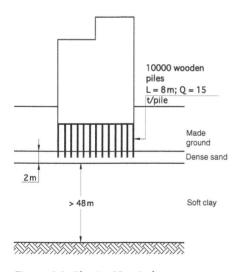

10000 wooden piles
L = 8 m; Q = 15 t/pile

Made ground

Dense sand

2 m

> 48 m

Soft clay

Test piles: Q = 30 t; w = 6mm

Settlement of the building

-   computed (from pile test and homogeneous subsoil): w = 80 mm
-   observed, end of construction: w = 40 to 100 mm
-   observed, 2 years after end of construction: w = 120 to 350 mm
-   final unknown but building demolished: w > 500 mm

Figure 5.2 Charity Hospital.

The non-linear behaviour even at low load level may be explained by two circumstances:

- The application of an axial load to a pile produces a concentration of shear stress in a thin layer of soil around the pile shaft; at low load level, the shear stress at the pile–soil interface is mobilized only in the upper part of the pile, always with relatively high values. Non-linear deformations are the consequence of such high local stress concentration.
- The shear strength at the pile–soil interface may be exceeded locally even at relatively low values of the applied load (Burland *et al.* 1977); as a consequence, even at relatively low load level local slip may occur.

Viggiani and Viggiani (2008) suggest evaluating the order of magnitude of the settlement w_s of an axially loaded single pile by the empirical expression:

$$w_S = \frac{d}{M} \frac{Q}{Q_{\text{lim}}} = \frac{d}{M} \frac{1}{FS} \tag{5.1}$$

where d is the pile diameter, Q is the applied load, Q_{lim} the bearing capacity and $FS = Q_{\text{lim}}/Q$ the factor of safety. Suggested values of M are listed in Table 5.2 as a function of the pile type and soil type. Eq. 5.1 with the values of M listed in Table 5.2 applies to $FS \geq 2.5$.

5.2.2 Load transfer curves

Let us consider (Figure 5.3) a pile subjected to an axial load Q, resisted by a distribution of shear stress τ at the pile–soil lateral interface and by a normal stress p at the pile base. The behaviour of the compressed pile is described by the following differential equation

$$EA \frac{\partial^2 w}{\partial z^2} = \Omega \tau(z) \tag{5.2}$$

where E is the Young modulus of the pile material, A the area of its section, Ω its perimeter. To complete the analysis, a relation between the vertical displacement w of a point at the pile–soil interface and the resulting shear stress τ is needed.

Table 5.2 Values of M, Eq. 5.1

Pile type	Soil type	M
Displacement	Cohesionless	80
	Cohesive	120
Small displacement (driven H or tube; large stem auger piles)	Cohesionless	50
	Cohesive	75
Replacement	Cohesionless	25
	Cohesive	40

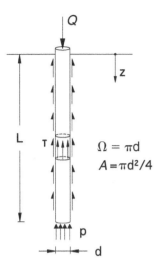

Figure 5.3 Scheme of a pile under axial vertical load.

Two possible strategies can be adopted:

- relating the displacement at a point only to the shear stress acting at the same point. Such an assumption, generally associated to the name of Winkler, assimilates the pile–soil connection to a set of mutually independent springs; the relation connecting the displacement w to the shear stress τ is known as "transfer curve" or t–z curve (with our notations, τ–w curve or p–w curve for the pile base);
- modelling the soil as a continuum, and imposing the compatibility of the displacement at the interface between the soil and the pile.

In the present paragraph, we shall explore the first strategy. In the case of a homogeneous soil, we can assume:

$$\tau = kw \tag{5.3}$$

where τ is the vertical shearing stress acting over a circumferential section of the pile and $k\,[FL^{-3}]$ is the spring constant applying to the soil alongside the pile. In other words, the transfer curve is assumed to be a straight line. The total force acting on the area Ωdz is:

$$\tau \Omega dz = \Omega kwdz$$

From Eq. 5.2 one obtains:

$$EA\frac{\partial^2 w}{\partial z^2} - \Omega k\,w = 0$$

Defining a characteristic length:

$$\lambda = \sqrt[2]{\frac{EA}{\Omega k}} \tag{5.4}$$

Eq. 5.2 becomes:

$$\frac{\partial^2 w}{\partial z^2} - \frac{w}{\lambda^2} = 0 \tag{5.5}$$

whose solution is:

$$w = C_1 e^{-\frac{z}{\lambda}} + C_2 e^{\frac{z}{\lambda}}$$

The boundary conditions are:

- at the top of the pile ($z = 0$; $-EA\dfrac{\partial w}{\partial z} = Q$):

$$Q = \frac{EA}{\lambda}(C_1 - C_2)$$

- at the tip of the pile ($z = L$; $P = pA = Akw_L$)

$$P = Ak\left(C_1 e^{-\frac{L}{\lambda}} + C_2 e^{\frac{L}{\lambda}} \right)$$

Closed form solutions for this and other simple cases are given by Scott (1981), Mylonakis and Gazetas (1998) and Salgado (2008).

A better approximation of the actual pile behaviour may be obtained considering that the actual relation between lateral stress and settlement is non-linear, and adopting non-linear empirical transfer curves instead of Eq. 5.3. As recalled above, load transfer curves are usually referred in the literature as t–z curves; with our notations they have to be called τ–w and p–w curves. They are obtained from load tests on piles with the shaft instrumented for the measurement of axial strain with depth (see Figure 7.13). The curves are usually reported in dimensionless form, with the displacement w normalized to the pile diameter d and shear stress τ or base normal stress p normalized to their ultimate value.

Typical transfer curves for side and base resistance of bored piles in clay and in sand are reported in Figures 5.4 and 5.5 (Reese and O'Neill 1988). The secant modulus of a τ–w curve gives the variable value of k making possible the solution of Eq. 5.5 by numerical methods (Figure 5.6). For instance, in terms of finite differences, it is possible to express the derivative of the function $w(z)$ at a point i as follows (Figure 5.7):

$$\frac{\partial^2 w}{\partial z^2} = \frac{w_{i+1} - 2w_i + w_{i-1}}{\Delta^2}$$

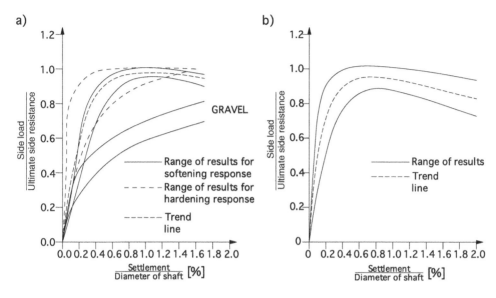

Figure 5.4 Transfer curves of side resistance for cohesionless (a) and cohesive (b) soils.

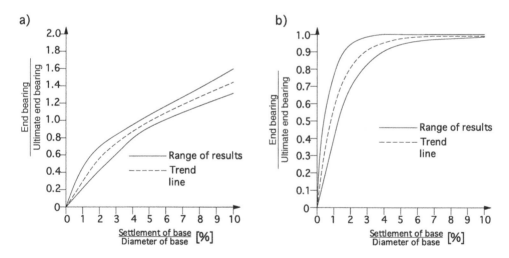

Figure 5.5 Transfer curves of base resistance for cohesionless (a) and cohesive (b) soils.

where $\Delta = L/n$. Eq. 5.5 becomes:

$$w_i = \frac{w_{i+1} + w_{i-1}}{2 + \left(\dfrac{1}{n^2} \dfrac{L}{\lambda_i} \right)} \tag{5.6}$$

where λ_i is obtained with the value of k of the soil surrounding point i; this allows an easy consideration of stratified soils. Eq. 5.6 may be written in the n points and, together with the proper boundary conditions, furnishes a system of linear equations

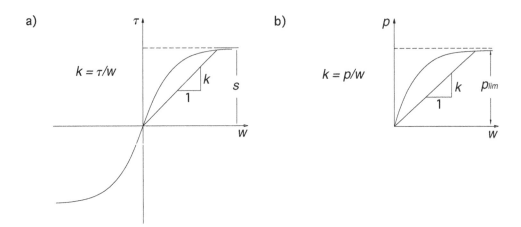

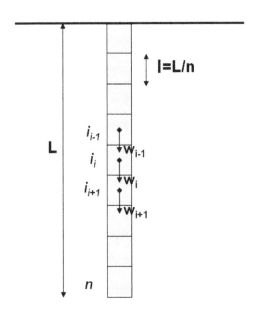

Figure 5.6 Typical transfer curves and definition of the secant modulus; (a) lateral and (b) base resistance.

Figure 5.7 Finite difference method for piles under axial load.

whose unknowns are the values of the displacement at the pile top and along the pile shaft. The system may be solved for each loading step, thus allowing the construction of the (non-linear) load–settlement relation for the pile top.

Randolph (1989) has developed a code named RATZ in which the shape of $\tau–w$ curves is that reported in Figure 5.8. The solution of the equation is obtained adopting the so-called explicit time approach (Cundall and Strack 1979), which involves the introduction of time as a variable, artificially in the case of static loading.

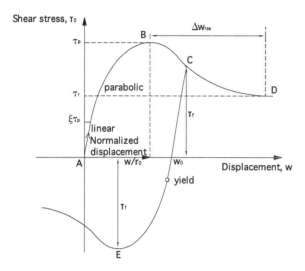

Figure 5.8 Transfer curves adopted in the code RATZ (after Randolph 1989).

In addition to monotonic and cyclic loading of the pile, RATZ may consider external soil movements causing downdrag, thermal strains in the pile, residual load developed during pile installation and matching of a measured load displacement response of the pile.

In Figure 5.9 two load–settlement curves obtained by pile load tests (Van Impe *et al.* 1998) are compared to the predictions performed by RATZ. The parameters needed have been evaluated fitting the experimental data. The agreement between predictions and experiments is rather satisfactory, and this occurs in the large majority of cases.

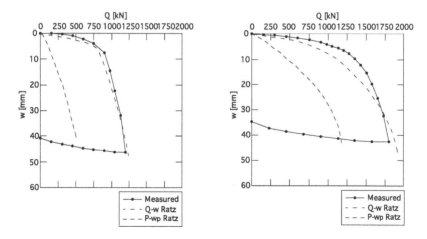

Figure 5.9 Comparison between pile load tests and prediction by RATZ (after Randolph 1989).

5.2.3 Elastic continuum: simplified analytical solution

Randolph and Wroth (1978) assume that the pile is immersed in an elastic half space, and analyse the interaction between the base of the pile and the elastic continuum separately from that between the latter and the lateral surface of the pile. The two solutions are then superimposed.

Let us consider first a rigid pile, transmitting the load to the surrounding elastic medium by lateral friction only. If τ is the value of the shear stress at the interface between the pile and the soil, at a radial distance r from the pile axis, to satisfy vertical equilibrium the shear stress τ_r is given by the expression:

$$\tau_r = \frac{\tau\, r_0}{r} \tag{5.7}$$

where $r_0 = d/2$ is the radius of the pile. The corresponding shear strain is

$$\gamma = \frac{\tau_r}{G} \tag{5.8}$$

where the shear modulus G is given by the expression:

$$G = \frac{E}{2(1+v)}$$

and v is the Poisson's ratio of the elastic medium. It can be assumed, as a first approximation, that:

$$\gamma = \frac{\partial w}{\partial r} \tag{5.9}$$

From Eqs 5.8 and 5.9 one obtains:

$$w(r) = \int_{r_0}^{r_m} \frac{\tau\, r_0}{Gr}\, dr = \frac{\tau\, r_0}{G} \ln\left(\frac{r_m}{r}\right)$$

where r_m is the radial distance at which the deformations are negligible ("magic radius" or extinction distance). The settlement w_S at the pile–soil interface $(r = r_0)$ is then:

$$w_S = \zeta \frac{\tau\, r_0}{G}$$

where $\zeta = \ln(r_m/r_0)$ is generally in the range from three to five.

The total load acting on the pile is given by $S = 2\pi r_0 L\tau$, and hence the ratio between the load and the settlement, or axial stiffness of the pile, is given by:

$$\frac{S}{w_S} = \frac{2\pi L \bar{G}}{\zeta} \tag{5.10}$$

where $\bar{G}$ represents the average value of G over the depth L.

The settlement w_b of the pile base subjected to a load P is given by:

$$w_b = \frac{1-v}{4} \frac{P}{r_b G_b}$$

where r_b is the radius of the pile base and G_b the shear modulus below the base of the pile. The related stiffness is:

$$\frac{P}{w_b} = \frac{4}{1-v} r_b G_b \tag{5.11}$$

If $Q = P + S$ is the total load acting on the pile, and w is the related settlement, Eqs 5.10 and 5.11 give:

$$\frac{Q}{w} = \frac{P}{w_b} + \frac{S}{w_S}$$

This equation may be expressed in the following dimensionless form:

$$\frac{Q}{w r_0 G_L} = \frac{4 r_b G_b}{(1-v) r_0 G_L} + \frac{2\pi L \bar{G}}{r_0 G_L} \tag{5.12}$$

where G_L represents the shear modulus at the depth L. This expression may be used to evaluate the settlement w of the pile subjected to a load Q.

Randolph and Wroth (1978) have generalized Eq. 5.12. Their results can be expressed as:

$$w = \frac{Q}{EL} I_w \tag{5.13a}$$

where:

$$I_w = \frac{2L(1+v)}{r_o} \frac{1 + \dfrac{4}{1-v} \dfrac{\eta}{\xi} \dfrac{tgh(\mu L)}{\mu L} \dfrac{L}{r_0}}{\dfrac{4}{1-v} \dfrac{\eta}{\xi} + \dfrac{2\pi\rho}{\zeta} \dfrac{tgh(\mu L)}{\mu L} \dfrac{L}{r_0}} \tag{5.13b}$$

in which the meaning of the symbols is as follows:

$\eta = \dfrac{r_b}{r_0}$ ($\eta = 1$ for cylindrical piles, $\eta > 1$ for piles with enlarged base)

$\xi = \dfrac{G_L}{G_b}$ relevant for piles crossing a layer with modulus G_L and point bearing on a half space with modulus G_b

$\rho = \dfrac{\bar{G}}{G_L}$ for piles crossing soils with variable stiffness

$$\lambda = \frac{E_p}{E_L} \quad \text{pile–soil relative stiffness}$$

$$\mu L = \sqrt{\frac{2}{\zeta \lambda} \frac{L}{r_0}}$$

According to the authors, parameter ζ may be expressed as follows:

$$\zeta = \ln\left\{\left[0.25 + \langle 2.5\rho(1-v) - 0.25\rangle \xi\right]\frac{L}{r_0}\right\} \tag{5.14}$$

Some results obtained by Eq. 5.9 are reported in Figure 5.10. In the case of piles floating in a homogeneous elastic half space ($\xi=1$) the settlement decreases with increasing stiffness $k = E_p/E_s$ and with decreasing slenderness L/d of the piles. The trend is typical of all soil–structure interaction problems; with increasing stiffness, the settlement merges into an asymptotic value, corresponding to a rigid pile. In the case of piles point bearing on a stiffer soil ($\xi=0.01$), as it was to be expected, the settlement is slightly smaller than that of the homogeneous half space.

5.2.4 Elastic continuum; solutions by BEM

The first application of the Boundary Element Method (BEM) to piles was published by Poulos and Davis (1968) and referred to a rigid pile while the soil was modelled as a linearly elastic, homogeneous and isotropic half space. Later on, quite a number of parametric studies have considered factors as finite stiffness of the pile, subsoil models other than the homogeneous half space (but, in any case, keeping the hypothesis of linear elasticity) and the occurrence of slips at the pile–soil interface.

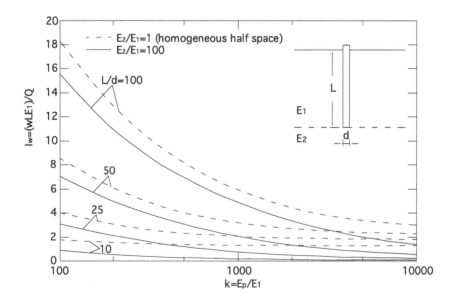

Figure 5.10 Some results by the application of the analytical approach (Eq. 5.13a).

A broad synthesis of these investigations is reported in the book by Poulos and Davis (1980). A short description of the method is reported.

Referring to Figure 5.11b, the distribution of shear stress acting at the lateral pile–soil interface is approximated by a finite number n of values, each acting on a cylindrical element of the pile of length $l = L/n$; let us call τ_i the value of shear stress acting on the element i, and p the uniform normal stress at the pile base.

The compatibility is imposed by equating the vertical displacement of the points at the pile–soil interface, at mid-height of each element, considered as belonging to the pile and to the soil.

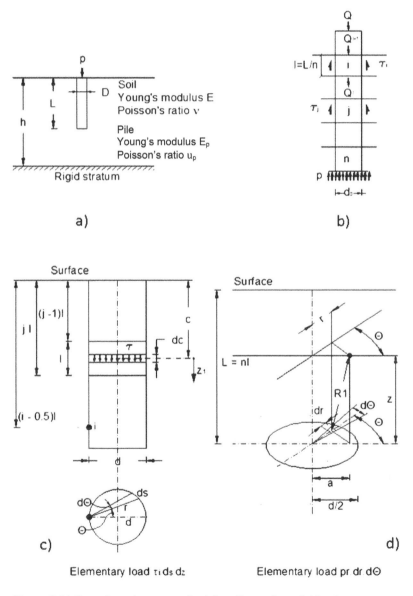

Figure 5.11 Boundary element method for piles under axial load.

Calling Q_{i-1} and Q_i the values of the axial force acting respectively on the upper and lower face of the element i, the compression Δw_i of the element may be expressed as:

$$\Delta w_i = \frac{4l}{\pi d^2 E_p} \frac{Q_{i-1} + Q_i}{2} \tag{5.15a}$$

The compression of the upper half of the element is:

$$\Delta w_i^U = \frac{4l}{\pi d^2 E_p} \frac{3Q_{i-1} + Q_i}{8} \tag{5.15b}$$

and that of the lower half:

$$\Delta w_i^L = \frac{4l}{\pi d^2 E_p} \frac{Q_{i-1} + 3Q_i}{8} \tag{5.15c}$$

It is easy to check that the Eq. 5.15a is equal to the sum of 5.15b + 5.15c.

If w is the displacement of the pile top, the displacement w_i of the midpoint of element i may be expressed as:

$$w_i = w - \sum_{j=1}^{i} \Delta w_j + \Delta w_i^L$$

or by the equivalent:

$$w_i = w - \sum_{j=1}^{i-1} \Delta w_j + \Delta w_i^U$$

By using Eqs 5.15, both these expressions give:

$$w_i = w - \frac{2l}{\pi d^2 E_p} \left\{ \sum_{j=1}^{i} \left(Q_{j-1} + Q_j \right) - \frac{Q_{j-1} + 3Q_j}{4} \right\} \tag{5.16a}$$

and, at the pile base:

$$w_b = w - \frac{2l}{\pi d^2 E_p} \sum_{j=1}^{n} \left(Q_{j-1} + Q_j \right) \tag{5.16b}$$

By vertical equilibrium one obtains:

$$Q_i = Q - \pi dl \sum_{j=1}^{i} \tau_j \tag{5.17}$$

Substitution of Eq. 5.17 into Eq. 5.16 gives the expressions of the displacements of the points belonging to the pile as a linear function of the $n+1$ unknown τ_i and of w.

The vertical displacement δw_{ij} of a point i belonging to the soil as a function of the shear stress τ_j on the element j may be expressed as:

$$\Delta w_{ij} = \frac{d}{E}\tau_j I_{ij}$$

where the influence coefficient I_{ij} is obtained by the Mindlin (1936) solution giving the state of stress and strain in an elastic half space loaded by a concentrated force acting in a vertical direction in a point of the half space. Making use of the so-called Steinbrenner approximation, the Mindlin solution may be extended to cases other than the homogeneous half space (layer of finite thickness H resting on a rigid substratum; Gibson's half space, characterized by a modulus linearly increasing with depth; layered medium etc.). Mindlin's solution is integrated numerically to obtain the effect of the shear stress τ_i uniformly distributed over the lateral surface of the cylindrical element j (Figure 5.11c) and of the normal stress p uniformly distributed at the pile base (Figure 5.11d).

Summing the effects of the various elements one obtains the expressions of the displacement of the points belonging to the soil:

$$w_i = \frac{d}{E}\left(\sum_{j=1}^{n}\tau_j I_{ij} + I_{ib}p\right) \tag{5.18a}$$

and, at the pile base:

$$w_b = \frac{d}{E}\left(\sum_{j=1}^{n}\tau_j I_{bj} + I_{bb}p\right) \tag{5.18b}$$

The unknown displacements are again linear functions of the $n+1$ unknowns τ_i and p.

The compatibility is imposed by equating the displacements given by Eqs 5.16 and 5.18, obtaining $n+1$ linear equations. A further equation is obtained by the vertical equilibrium:

$$\frac{\pi d^2}{4}p + \pi dl\sum_{i=1}^{n}\tau_i = Q$$

Summing up, there are $n+2$ equations in the $n+2$ unknowns τ_i, p and w.

After solving the system of linear equations, the settlement w of the top of a single pile subjected to an axial load Q may be expressed as:

$$w = \frac{I_w Q}{EL} = w_1 Q \tag{5.19}$$

where w_1 is the axial compliance of the pile and I_w is a dimensionless influence coefficient, function of the parameters L/d, v, $K = E_p/E$ and of the selected elastic subsoil model (the available subsoil models are: homogeneous half space; layer of finite thickness H resting on a rigid substratum; Gibson's half space, characterized by a modulus linearly increasing with depth; horizontally layered medium etc.); v is the Poisson's ratio of the elastic medium modelling the subsoil.

There are quite a number of computer programs available for the computation of the settlement of a single pile. The program SINGPALO (Mandolini and Viggiani 1997) is based on the following assumptions:

- horizontally layered elastic soil; the layering beneath the pile point is accounted for by the so-called Steinbrenner approximation, as described by Poulos and Davis (1980); the layers crossed by the pile shaft are treated in the same way by an approximate application of the reciprocal theorem;
- pile with stepwise or continually variable section;
- compatibility of both horizontal and vertical displacements;
- slip at the pile–soil interface once a limiting shear stress s has been attained; the latter can be either cohesive or equal to the horizontal stress times a friction coefficient.

Some of the results obtained by SINGPALO are reported in Figure 5.12. The results are very similar to those obtained by the approximate analytical solution by Randolph and Wroth (1978) and reported in Figure 5.10. Plenty of similar results may be found in the treatise by Poulos and Davis (1980).

From a practical point of view, the availability of elastic solutions does not improve significantly our ability to predict the settlement of a single pile; as we have

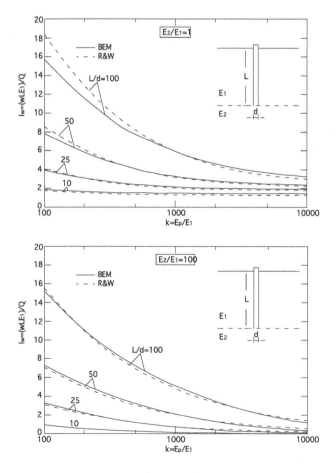

Figure 5.12 Comparison between the results by SINGPALO (BEM) and Randolph and Wroth (1978).

seen, in fact, the load–settlement behaviour of a pile is markedly non-linear and the use of an expression such as Eq. 5.19 requires the evaluation of an equivalent modulus which is far from straightforward. The value of the solutions, however, is to be found in the possibility of analysing the interaction among piles in a group, and hence evaluating the settlement of pile groups. From this point of view, modelling the subsoil as a continuum, even the very simple linearly elastic continuum, represents a significant conceptual improvement over the transfer curve approach.

5.2.5 Solutions by FEM

The settlement of a pile subjected to a vertical axial load may be analysed by finite elements, taking advantage of the axial symmetry of the problem. From a numerical point of view, FEM is less efficient than BEM since the former implies the discretization of the whole continuum, while the latter requires only the discretization of the boundary between the pile and the soil. FEM, however, is much more general allowing consideration of a non-homogeneous soil profile, of any mechanical property of the pile–soil interface and even non-linear constitutive modelling of the soil.

One of the first applications of FEM to the analysis of piles was reported by Ellison *et al.* (1971); they developed an axi-symmetric model to back analyse five load tests on bored piles in London Clay. The details of the model are reported in Figure 5.13a; spring elements were provided at the interface between the pile shaft and the soil to allow for slip and tension cracks, and beneath the pile tip to allow for direct calculation of the tip load. Figure 5.13b shows the pattern of principal stresses in the vicinity of the pile tip; the model revealed large changes in principal stresses direction and magnitude and the occurrence of tension cracks.

Potts and Zdravkovicz (2001) review some of the pitfalls occurring in the selection of the most suited constitutive model of the soil. Figure 5.14a shows the axi-symmetric finite element model used to predict the response of a pile 1 m in diameter and 20 m long. The Mohr Coulomb model was used with an angle of dilatancy Ψ ranging between zero and the friction angle φ of the soil. The mobilization of the shaft and base resistance for the two analyses are shown in Figure 5.14b. For $\Psi = \varphi$ both the base and the shaft show no sign of reaching a limiting value, because of the kinematically constrained nature of the problem. If the dilatancy angle is set to zero, on the contrary, a realistic prediction of the pile response is obtained. These results, as many others, confirm the sensitivity of the output of numerical analyses to the details of the constitutive model.

At present, the use of FEM in design seems to be still out of the question, mainly because of the difficulties connected with a proper characterization of the soils; the value of FEM analyses is to be found in the development of parametric studies and benchmark solutions.

5.3 Settlement of pile groups

5.3.1 Empirical methods

The settlement of a pile belonging to a group, and hence the average settlement of the group, is always larger than the settlement of a single pile subjected to a load

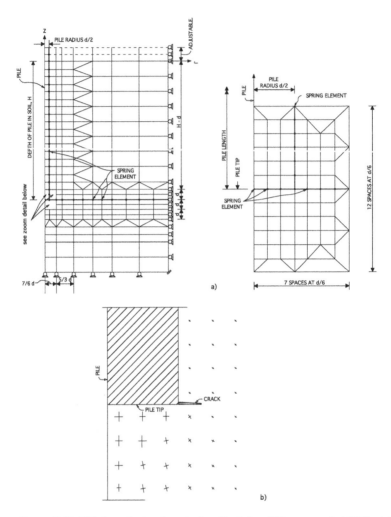

Figure 5.13 FEM analysis of a single pile (after Ellison *et al.* 1971): (a) finite element mesh; (b) principal stress in the vicinity of the pile tip.

equal to the average load per pile of the group. This is an effect of the interaction among the piles via the surrounding soil, as schematically shown in Figure 5.15.

The average settlement w_g of a group of n piles may be expressed as a function of the settlement of a single pile subjected to a load equal to the average load per pile in the group:

$$w_g = R_S w_s = nR_G w_s \tag{5.20}$$

where w_s is the settlement of a single pile under the average working load Q/n of the group (Q = total load applied to the foundation), R_S is an amplification factor named group settlement ratio, originally introduced by Skempton *et al.* (1953) and quantifying the effects of the interaction between piles, and $R_G = R_S/n$ is the group reduction factor. It may be shown that $1 \le R_S \le n$ and hence $1/n \le R_G \le 1$.

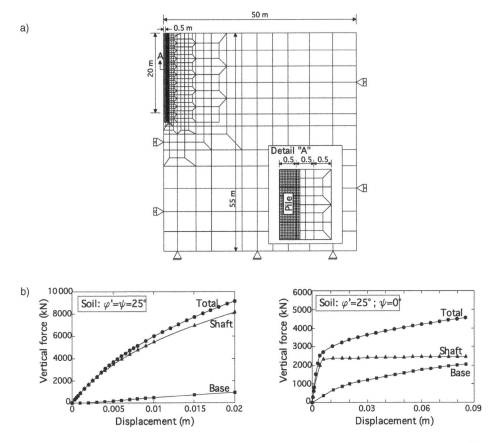

Figure 5.14 FEM analysis of a single pile (after Potts and Zdravkowictz 2001): (a) finite element mesh; (b) load–settlement curves for $\varphi' = \Psi$ and for $\Psi = 0$.

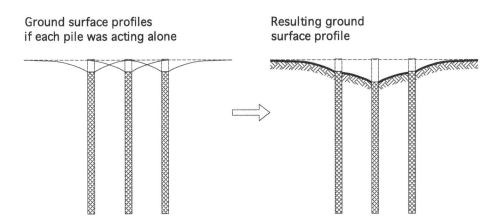

Ground surface profiles if each pile was acting alone

Resulting ground surface profile

Figure 5.15 Superposition of the deformation fields in a pile group.

Empirical expressions for R_S had been suggested by Skempton *et al.* (1953), Meyerhof (1959) and Vesic (1968). Mandolini *et al.* (2005) have collected 63 well-documented case histories of the settlement of piled foundations for which the settlement records and the results of load tests on single piles are available. A wide range of pile types (driven, bored, CFA) assembled in a variety of geometrical configurations ($4 \leq n \leq 6500$; $2 \leq s/d \leq 8$; $13 \leq L/d \leq 126$) and regarding very different soils (clayey to sandy soils, stratified, saturated or not etc.) are included.

The experimental values of $R_S = w_g/w_s$ are reported in Figure 5.16 as a function of the size of the group, expressed by the pile number n. It is evident that R_S is always larger than unity, and increases with increasing n; the scatter of the data is due to the influence of factors as the actual shape of the group, the nature of the subsoil, the spacing, length and relative stiffness of the piles. A more significant geometrical parameter characterizing the group is the aspect ratio $R = \sqrt{\dfrac{ns}{L}}$, introduced by Randolph and Clancy (1993). Plotting the group reduction factor R_G as a function of R, one gets Figure 5.17.

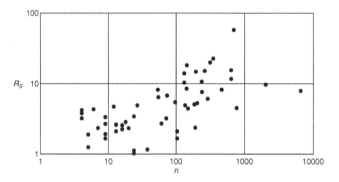

Figure 5.16 Increase of the group settlement ratio R_S with the number of piles in a group.

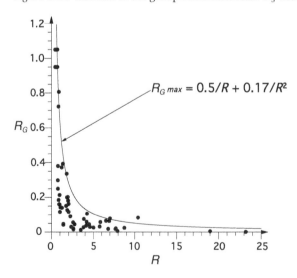

Figure 5.17 Relationship between $R_{G\max}$ and R.

On this empirical basis the following expressions for the upper limit R_{Smax} and the best estimate of R_S, as a function of R are obtained:

$$R_{Smax} = \frac{w_{gmax}}{w_s} = \frac{0,50}{R} \cdot \left(1 + \frac{1}{3R}\right) \cdot n \qquad (5.21)$$

$$R_S = \frac{w_g}{w_s} = 0,29 \cdot n \cdot R^{-1,35} \qquad (5.22)$$

Eqs 5.21 and 5.22 may be useful for preliminary evaluations. The value of w_s is best obtained by full-scale load test on single pile; since the results of load tests are rarely available at the design stage, w_s can be evaluated alternatively by the procedures reported in §5.2.

5.3.2 Equivalent raft and equivalent pier

An evaluation of the settlement of a pile group can be obtained assimilating the group to an equivalent shallow foundation (an equivalent raft) or to a single equivalent pile or pier, in order to take advantage of the procedures for the computation of settlement of a shallow foundation or single pile. Both methods have been used in a number of slightly different formulations, specially the former, the most usual.

In the method of the equivalent raft, the pile foundation is assimilated to a raft placed at a certain depth, with a size in plan established by criteria as those illustrated in Figure 5.18.

The suggestions of different authors (Tomlinson 1987, 1994; Van Impe 1991; Viggiani 1993; Randolph 1994) may be rather different and leave ample room to the designer judgement. In Figure 5.18 the indication by Tomlinson (1994) is reported.

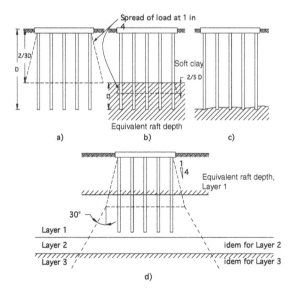

Figure 5.18 Equivalent raft.

In cohesive soils, the settlement may be computed by the classical method of Skempton and Bjerrum (1957); the undrained settlement is evaluated by the design charts of Christian and Carrier (1978) for homogeneous soil and of Butler (1974) or Esposito *et al.* (1975) for soils whose stiffness increases with depth. Such charts are reported in Figures 5.19 and 5.20. Viggiani (1993) suggests correlating the values of the undrained modulus E_u to the undrained shear strength c_u; values of E_u/c_u are reported in Table 5.1.

In cohesionless soils, methods as Burland and Burbidge (1985) or Schmertmann *et al.* (1978), based respectively on SPT and CPT results, may be adopted. In any case, calculations are simple and do not require the use of a computer.

The method of the equivalent pier has been suggested by Poulos and Davis (1980) and consists in substituting a single equivalent pier to the pile group and interposed soil. The equivalent pier may have an equivalent length L_e and the same size in plan of the group, or the same length and an equivalent diameter d_e. Randolph (1994) suggests this latter criterion, assuming:

$$d_e = \sqrt{\frac{4}{\pi} A_g} = 1,13\sqrt{A_g}$$

where A_g represents the area of the pile group. The modulus E_{eq} of the equivalent pier is given by:

$$E_{eq} = E + \left(E_p - E\right)\left(\frac{A_p}{A_g}\right)$$

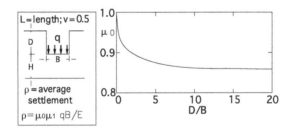

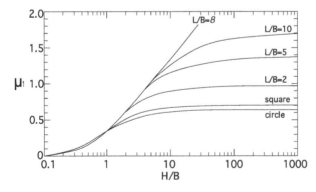

Figure 5.19 Solutions for a deep loaded area on an elastic layer (after Christian and Carrier 1978).

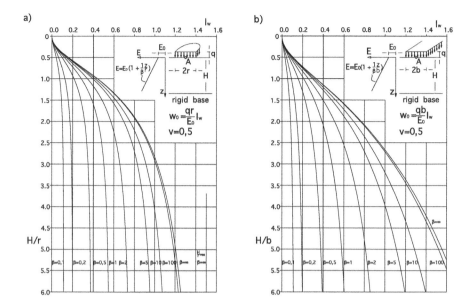

Figure 5.20 Chart for the evaluation of immediate settlement in a soil with the stiffness line-arly increasing with depth: (a) circular foundation; (b) long strip (source: Esposito *et al.* 1975).

where E is the modulus of the soil, E_p that of the pile material and $A_p = n\pi d^2/4$ is the area occupied by the piles. The settlement of the equivalent pier is then evaluated by the methods reported in §5.2.

The equivalent raft method is to be preferred for large pile groups, or more pre-cisely for groups where the breadth B is larger than the length of the piles L, while the equivalent pier method is more suited when $L > B$ (Figure 5.21). Randolph (1994) suggests using the former for values of the aspect ratio $R > 4$, and the latter for $R < 2$; in intermediate cases, judgement must be exerted.

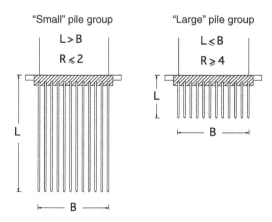

Figure 5.21 Small and large pile groups.

Both methods, however, are rather subjective and may be properly modified to adapt them to the particular conditions of the case under investigation; for these reasons, they are markedly dependent on the designer's judgement. This point may be illustrated by an example.

Committee TC18 (*Pile Foundations*) of the ISSMGE carried out an assessment of the methods for the prediction of the settlement of pile foundations, by circulating among a number of leading geotechnical engineers the subsoil properties and the geometrical and load data of some piled foundations whose settlement had been measured, but was not known to the participants. One of the foundations was the pier Number Seven of the bridge over the Garigliano river (Mandolini *et al.* 2005). Van Impe and Lungu (1996) report the results of the evaluation. One of the most successful predictions was carried out by Tomlinson using the method of equivalent raft, in spite of the value of $R = 1.2$; the method had been adapted to the case by adopting different schemes for the immediate and consolidation settlement (Figure 5.22).

5.3.3 Elastic continuum

The Boundary Element Method introduced in §5.2.4 for the single pile may be easily extended to model a pile group. Each of the piles in the group is discretized as in §5.2.4, and global equilibrium and compatibility of displacements are imposed, considering the mutual interaction of all the elements of all the piles. The computational resources

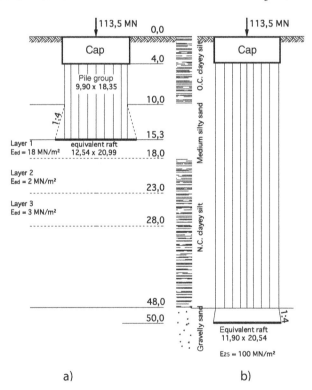

a)   b)

Figure 5.22 Prediction of the settlement of pier Number 7 of the bridge over the Garigliano by the method of equivalent raft: scheme for (a) the prediction of the immediate settlement and (b) for the final settlement.

required by this procedure (the so-called "complete BEM"), however, become very large as soon as the number of piles in the group exceeds some tens, especially if a step-wise incremental procedure is adopted to simulate a non-linear behaviour.

One of the procedures available to overcome such a difficulty is the superposition of interaction coefficients. To demonstrate the method, let us consider a group formed by two identical piles i and j, at a spacing s axis to axis, and subjected to the loads Q_i and Q_j. After having solved the problem by BEM, the settlement of one of the piles may be expressed as:

$$w_i = w_1 \left(Q_i \alpha_{ii} + Q_j \alpha_{ij} \right)$$

in which $\alpha_{ii} = 1$ and α_{ij} is the interaction coefficient between piles i and j. The interaction coefficient is a function of the same quantities of I_w, (Eq. 5.14) plus the ratio s/d between the spacing and the diameter. When s/d approaches zero, α_{ij} approaches unity; when s/d tends to infinity, α_{ij} tends to vanish. Values of α_{ij} have been obtained by BEM for different subsoil models (Poulos 1968; Banerjee and Driscoll 1976; Caputo and Viggiani 1984; Bilotta *et al.* 1991). Randolph and Wroth (1979) have provided approximate analytical expressions. An extinction distance r_m is sometimes adopted, beyond which α_{ij} is put equal to zero. For instance, Randolph and Wroth suggest to assume for r_m the value given by Eq. 5.14.

Some values of the interaction coefficients are reported in Figure 5.23. It may be seen that, for a homogeneous half space, the more the piles are long and deformable, the more they are interactive.

Mandolini and Viggiani (1997) suggest determining by BEM some values of the interaction coefficients for the case under consideration at various distances between the two piles, and then fitting it by closed form expressions, thus enabling a simpler computer analysis of group settlement behaviour. For example, the following expressions have been suggested:

$$\alpha = A \left(\frac{s}{d} \right)^B$$

$$\alpha = \left\{ C + D \ln \left(\frac{s}{d} \right) \right\}$$

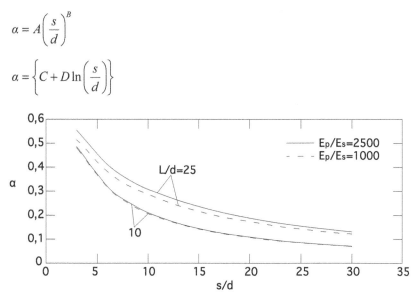

Figure 5.23 Interaction coefficients for piles embedded in an elastic half space.

where A, B, C, D are fitting parameters. Alternatively, simple interpolations between calculated values may be adopted.

Once obtained, the interaction coefficients can be used to simulate the interaction of a pile as a whole with another pile within the group, assuming that it is not influenced by the presence of the other piles and hence the interaction coefficients obtained for the pair of piles are still valid. The superposition of the interaction coefficients obtained for the pair of piles neglects, in the analysis of each pair of piles within the group, the stiffening effect of the other piles; the interaction, therefore, is somewhat overestimated. The error involved, however, is small as shown by the results reported in Table 5.3.

In the method of the interaction coefficients the settlement of a pile i within a group may be expressed as:

$$w_i = \sum_{j=1}^{n} w_{1i} Q_j \alpha_{ij} \tag{5.23}$$

The analysis of the group is very simple in two cases: when the loads Q_i on the piles are known (and possibly equal), and when the settlement of the group is constant or linearly variable. The two cases are commonly associated to an infinitely flexible and respectively to an infinitely rigid cap. In the first case Eq. 5.23 can be used directly to

Table 5.3 Values of R_G and load distribution on piles

	Group reduction factor R_G			
Group type	2×2	3×3	4×4	5×5
Interaction coefficients (Poulos 1968)	0.672	0.541	0.460	0.403
Complete BEM (Butterfield and Banerjee 1971)	0.665	0.550	0.456	0.396

		Load distribution on piles	
Group type	*Pile*	\multicolumn{2}{c}{Q/Q_{av}}
		Inter. coefficients	*Complete BEM*
3×3	Corner	1.52	1.51
	Side	0.74	0.75
	Centre	−0.05	−0.06
4×4	Corner	2.02	2.02
	Side	0.96	0.96
	Centre	0.05	0.04
5×5	Corner	2.58	2.52
	Side	1.18	1.19
	Side	1.16	1.16
	Intermediate	0.01	0.05
	Intermediate	0.01	0.11
	Centre	0.19	0.10

evaluate the settlement of any pile in the group. In the second case (Figure 5.24) it has to be associated to the equilibrium conditions:

$$\sum_{i=1}^{n} Q_i = Q \qquad \sum_{i=1}^{n} Q_i x_i = Q e_x \qquad \sum_{i=1}^{n} Q_i y_i = Q e_y \qquad (5.24)$$

and to the compatibility conditions which, assuming identical piles ($w_{1i} = w_{1j} = w_1$), is:

$$w_i = w_1 \sum_{j=1}^{n} Q_j a_{ij} = w_0 + a_x y_i + a_y x_i \qquad (5.25)$$

in which the vertical displacement of the centre w_0 and the two rotations a_x and a_y define the rigid motion of the pile cap. Eqs 5.24 and 5.25 form a system of $(n+3)$ linear equations in the $(n+3)$ unknowns Q_i, w_o, a_x and a_y.

The values of the interaction coefficients are influenced by the subsoil profile; in general, the interactivity increases for the profiles of decreasing stiffness with depth (Figure 5.25) and vice versa. Some examples of the dependence of the interaction

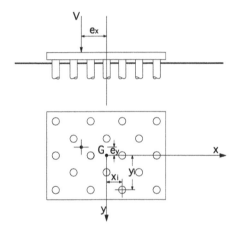

Figure 5.24 Group of piles with rigid cap.

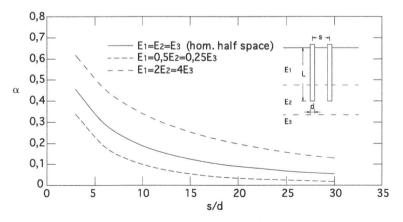

Figure 5.25 Influence of the subsoil profile on the values of interaction coefficients.

coefficients on the subsoil model are provided by Poulos (2008), who emphasizes that misleading results can arise from imprudent application of group settlement analysis.

The evaluation of the elastic properties of the subsoil is the most uncertain step of the prediction of the settlement of a pile group; to overcome such a difficulty, a number of authors (among others, Poulos 1972b; Mandolini and Viggiani 1997) have suggested of backfiguring such properties by the axial stiffness of a single pile, as obtained for instance in a load test. The topic is discussed in detail in the next section, 5.3.4; it is to be pointed out, however, that such a suggestion could induce engineers to limit the design investigations to a load test on a prototype pile. This would be a dangerous practice, as pointed out by Russo and Viggiani (1997), because of the dependence of the interaction coefficients on the subsoil profile. The already-quoted case history of the Charity Hospital in New Orleans (see Figure 5.2) is a well-known example; other examples of such experience are reported by Golder and Osler (1968) and Peaker (1984); further elements are those presented in Figure 5.25.

Poulos (1968) has shown that the average settlement of a pile group is but slightly affected by the assumption of flexible or rigid cap. In the majority of the practical applications it is suggested that the load on each pile be evaluated by the associated influence area, and Eq. 5.24 is applied to a pile in an intermediate position between the centre and the side of the group. The analogy with the concept of characteristic point of a shallow foundation is evident.

Figure 5.26 reports some results obtained by the program GRUPPALO (Mandolini 1994; Viggiani 1998) for different group configurations (square, rectangular and circular), subsoil models (half space and layer of finite thickness resting on a rigid base) and pile slenderness (L/D = 25–100). It may be seen that the settlement of the group increases with the number of piles; since the influence of the group shape is negligible and the spacing of the group in Figure 5.26 is constant and equal to 3*d*, it may also be concluded that the group settlement increases with increasing breadth of the group. The values of the group settlement ratio decrease with decreasing the thickness of the deformable layer *H*; the elastic half space (H/L = ∞) is the model in which the interaction between the piles is maximum, and hence the group settlement ratio has the highest values.

Figure 5.27 (Mandolini and Viggiani 1997) provides some elements to evaluate the overall accuracy of the code GRUPPALO.

5.3.4 Evaluation of soil properties and implementation of the analysis

It has been already noted that the elastic properties of the soil, to be used in the analysis, are very difficult to evaluate because of the marked non-linearity of the stress-strain relation and the influence of pile installation. A number of authors have suggested utilizing to this aim the results of pile load tests. Such tests are sometimes available, even at the design stage, for important projects; should this not be the case, the load–settlement behaviour of a single pile can be simulated, for instance, by the transfer curves approach.

The evaluation of the elastic parameters of the subsoil from the load–settlement relation of a load test on a pile is an uncertain and somewhat subjective exercise. In

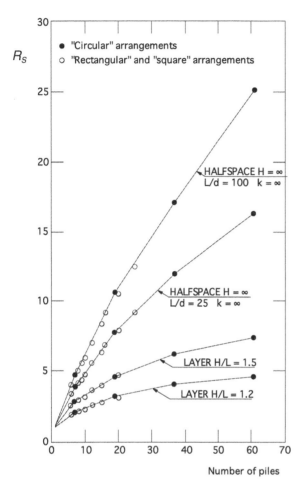

Figure 5.26 Increase of the group settlement ratio R_S with the number of piles.

order to reduce the uncertainties, a standard procedure has been developed, involving the use of the results of any traditional subsoil investigation. The procedure is described in detail by Viggiani (1998), and is summarized in the following (Figure 5.28). The results of all the available site and laboratory investigations are first used to develop a model of the subsoil, in which the geometry is adapted to a scheme of horizontal layering. The relative stiffness of the layers is also evaluated, such an evaluation being relatively easy on the basis of the results of laboratory tests, or site tests as CPT, SPT, DMT. The absolute values of the stiffness of the different layers are then fixed by fitting the results of an elastic analysis of the single pile, based on the previously developed subsoil model, to the load–settlement curve of the same pile obtained by a load test, or simulated with a suitable procedure. Once the subsoil model is fixed and the stiffness of each layer is established, the same model is used for the analysis of the piled foundation.

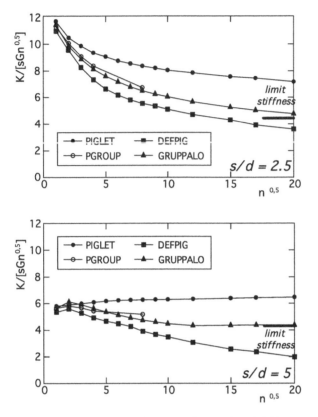

Figure 5.27 Comparison between the results of different programs for the evaluation of the settlement of a pile group.

Of course, the elastic analysis of the single pile produces a rectilinear load–settlement relation; it may be fitted either to the initial tangent of the actual non-linear load–settlement curve, or to the secant connecting the origin to the settlement under the average working load. The former choice may be properly considered a linearly elastic (LE) analysis; the latter is an elastic analysis based on a secant modulus (ES). The two procedures lead obviously to different results, and in some cases the difference may be significant. A comparison with the experimental evidence clarifies the meaning of the different analyses and helps selecting the one most suited.

Mandolini *et al.* (2005) report a comparison between the observed average settlement of 48 case histories and the predictions obtained by GRUPPALO. The majority of the analysed foundations had been designed according to a conventional capacity-based approach. As a consequence, their safety factor under the working load is rather high, and a simple linear analysis may be expected to be adequate for engineering purposes. Indeed the LE analysis, based on the moduli backfigured by the initial stiffness of the load test on single piles, gives a rather satisfactory agreement with the observed values in all these cases (Figure 5.29a, open dots).

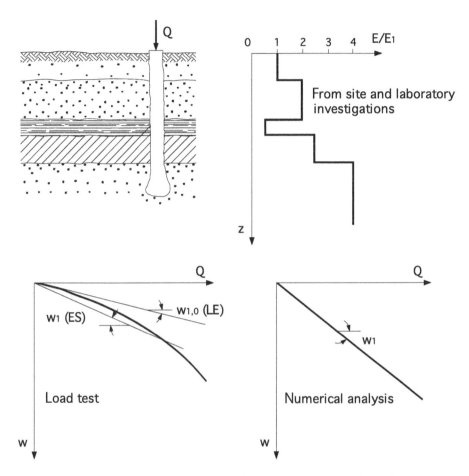

Figure 5.28 Evaluation of subsoil properties by back analysis of a pile load test.

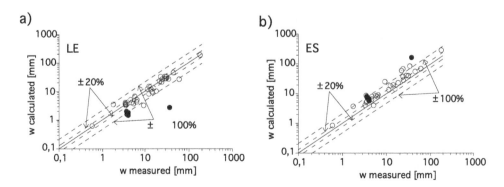

Figure 5.29 Comparison between predicted and measured settlement: (a) linearly elastic analysis with initial tangent modulus; (b) linearly elastic analysis with secant modulus.

There are, however, some cases in Figure 5.29a (full dots) referring to small pile groups constructed for research purposes and submitted to a load level close to failure (Brand *et al.* 1972; Briaud *et al.* 1989). For these cases LE analysis is less satisfactory, resulting in a substantial underestimation of the settlement; it appears that non-linearity plays a major role and hence its consideration is mandatory. The topic will be examined in §5.3.5.

The ES analysis (Figure 5.29b), on the contrary, substantially overestimates the observed settlement. It appears that the choice of performing an elastic analysis on the basis of some secant modulus, the most widespread and apparently the most reasonable one, is in fact rather misleading.

5.3.5 Non-linearity

The analysis of the load–settlement behaviour of a single pile, accounting for non-linearity, may be carried out by the transfer curve approach (§5.2.2). It is also possible to predict a non-linear response from elastic BEM solutions, by an incremental analysis imposing the condition that the shear stress at the lateral pile–soil interface and the normal stress at the pile base cannot exceed an ultimate value, which can be evaluated with the criteria presented in Chapter 4. In structural analysis such a procedure is known as Pushover Analysis.

This feature is implemented in the program SINGHYP (Russo 1996; Van Impe *et al.* 1998) which, in addition, introduces a hyperbolic load–settlement response at the pile base. Such a relation is completely defined by its initial tangent and its ultimate asymptotic value. The former of these quantities can be evaluated by elasticity theory, while the ultimate bearing capacity by the criteria presented in Chapter 4. Both evaluations belong to design routine.

Predictions by SINGHYP are compared in Figure 5.30 to the same load–settlement curves used in Figure 5.9; it may be seen that the agreement between prediction and experimental data is even more satisfactory than with the transfer curves approach.

As far as the group behaviour is concerned, once the non-linear behaviour of the single pile is defined, in principle there are no problems in describing the non-linear behaviour of the group by a stepwise incremental linear analysis in which the stiffness of each pile is updated as a function of the load level; iteration within each loading step may improve the accuracy of the results. It is obvious, however, that the computing resources and time needed make such an analysis long and cumbersome.

Caputo and Viggiani (1984) report some experimental data obtained by pile load tests in which, in addition to the settlement of the test pile, the settlement of unloaded adjacent piles had been measured. A sample of such data is reproduced in Figure 5.31. The load–settlement curves of the loaded (source) pile has the usual non-linear trend, while the curve relating the settlement of the unloaded (receiver) pile to the load acting on the loaded (source) test pile is very nearly linear. It appears that, at some distance from an axially loaded pile, the deformation of the soil is essentially linear, and this applies to the displacement of any unloaded pile within the deformation field of the source pile. Such a behaviour is consistent with the rapid decay of stress moving radially away from the pile–soil lateral interface, as argued by Cooke (1974) and Frank (1974) and expressed by Eq. 5.7.

In order to get a better insight into the phenomenon, let us consider two identical adjacent piles (Figure 5.32) and analyse their behaviour when pile 1 is subjected to

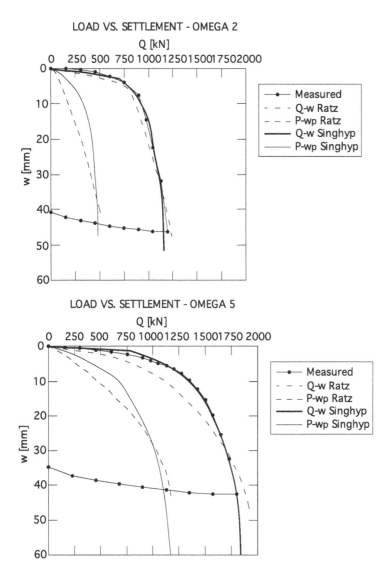

Figure 5.30 Comparison between pile load tests and prediction by SINGHYP and RATZ.

an axial load Q gradually increasing to the ultimate value Q_{lim} while pile 2 is kept load free. The analysis is carried out incrementally, allowing local slip to occur at the pile–soil interface when the interface stress reaches its limiting value. The results obtained are reported in terms of the ratio between the interaction factor α_{12} evaluated by the non-linear analysis and the corresponding value from a linearly elastic analysis, plotted against the ratio $Q/Q_{lim} = 1/FS$. As observed in the experiments of Figure 5.31, the interaction factor keeps constant up to value of the safety factor of the order of 2.5 – i.e. the value usually adopted in design – and even for FS as low as 1.5 the decrease of the interaction factor is less than 10%. In a way, such a result could have been foreseen on the basis of the so-called Saint Venant principle.

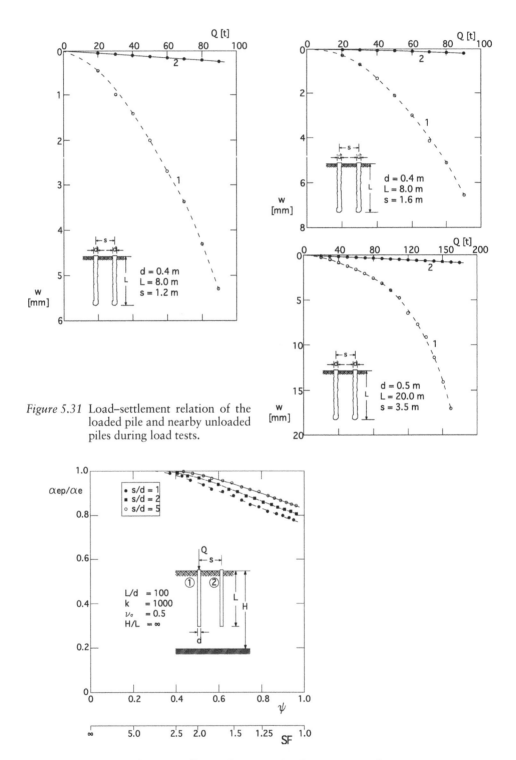

Figure 5.31 Load–settlement relation of the loaded pile and nearby unloaded piles during load tests.

Figure 5.32 Non-linearity effect in the interaction between two piles.

On the basis of these and similar results, Caputo and Viggiani (1984) suggested implementing the non-linear incremental analysis of a pile group by updating, at each load increment, the coefficient α_{ii} expressing the effect of the load Q_i on the loaded pile while keeping constant the coefficient α_{ij} expressing the effect of the load Q_i acting on the pile i on the settlement of the pile j. In other words, in the expression:

$$\{w_i\} = w_i\left[\alpha_{ij}\right]\{Q_j\}$$

the terms on the principal diagonal of the matrix of the interaction coefficients α_{ii} are updated while the terms off diagonal α_{ij} are kept constant, strongly decreasing the computational effort and improving the fidelity of the model. In this way, the non-linearity is concentrated at the pile–soil interface, while the interaction between the piles are treated as linear. It may be shown that this procedure is essentially equivalent to the suggestion by Randolph (1994), to estimate the group response on the basis of the initial small strain elastic stiffness and afterwards adding the plastic displacement due to the slip at the pile–soil interface.

The load–settlement relation of the loaded piles may be assimilated to a hyperbola (Chin 1970) whose equation is:

$$Q = \frac{w}{a+bw}$$

where $b = 1/Q_{\lim}$ is the inverse of the bearing capacity of the pile and $a = \left[\dfrac{\partial w}{\partial Q}\right]_{Q=0}$ is the inverse of the initial tangent stiffness of the pile ($a = w_1$). It may be shown that, with this assumption, the updating of the interaction coefficients may be done according to the expression:

$$\alpha_{ii} = \frac{Q_{\lim}}{Q_{\lim} - Q_i}$$

The procedure is incorporated in the code of GRUPPALO and has been adopted for the back analysis of the same case histories of Figure 5.28; the results obtained are plotted in Figure 5.33. The NL analysis, which essentially consists in adding the non-linear component of the settlement of the single pile to the settlement of the group, obtained as in the LE analysis, slightly improves the prediction of the average settlement in all the cases where the LE analysis was already successful. In the cases where the non-linearity plays a significant role, NL analysis significantly improves the prediction.

5.4 Differential settlement

As stated in §3.2, the design against serviceability limit state requires that the overall and differential settlement of a foundation be kept below the corresponding admissible values. While some suggested admissible values are listed in §3.3.3, the procedure for the prediction of the differential displacements of a pile foundation are reviewed herewith.

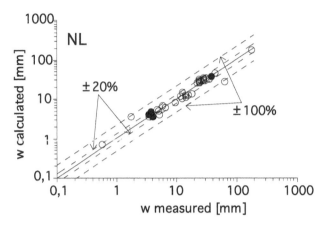

Figure 5.33 Comparison between predicted and measured settlement; non-linear analysis by NAPRA.

The expected maximum values of differential settlement, deflection and angular distortion, to be compared to the admissible values, are difficult to obtain as the result of a deterministic design analysis, being markedly dependent on random factors such as variability of soil properties, installation procedures and history of construction and loading. It is therefore useful, at least as a first step, to adopt empirical relations between the distortion parameters and some readily evaluable quantity.

Some experimental data concerning the maximum expected value of the differential settlement δ_{max} of pile foundations, expressed as a fraction of the average settlement w_g ($R_{D\,max} = \dfrac{\delta_{max}}{w_g}$), have been collected by Mandolini *et al.* (2005) and are reported in Figure 5.34.

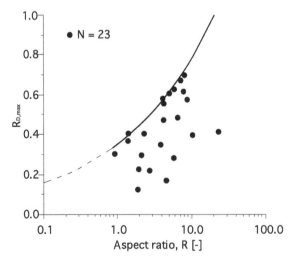

Figure 5.34 Relationship between $R_{D\,max}$ and R.

The empirical evidence on the relation between the maximum settlement w_{max}, that is readily evaluated, and the maximum expected value of the angular distortion β_{max}, including both shallow and pile foundations, is summarized in Figure 5.35 (Viggiani 1999); the upper envelope of the experimental data may be expressed as:

$$\beta_{max} = 10^{-4} w_{max} \, (mm)$$

A limited evidence relating specifically to pile foundations in clay (Grant *et al.* 1974) seems to indicate that the presence of piles reduces the maximum expected angular distortion for a given value of the maximum settlement to the value:

$$\beta_{max} = 6.2 \times 10^{-5} w_{max} \, (mm)$$

Such an effect had been predicted by Burland *et al.* (1977).

A deterministic procedure to evaluate an upper bound of differential settlement, deflection ratio and angular distortion is that of performing settlement calculations for the examined pile foundation in the hypothesis of flexible foundation, assuming that the load on each pile is known and obtained by the influence area of the pile. If the application of such a procedure results in unreasonably high distortion, the finite stiffness of the raft can be considered by the procedures that will be dealt with in Chapter 6.

As an example, the case history of two 90 m high towers recently built in eastern Naples area is presented (de Sanctis *et al.* 2002). They are founded on two adjacent piled rafts with 637 CFA piles, 0.6 m in diameter and 20 m in length, uniformly

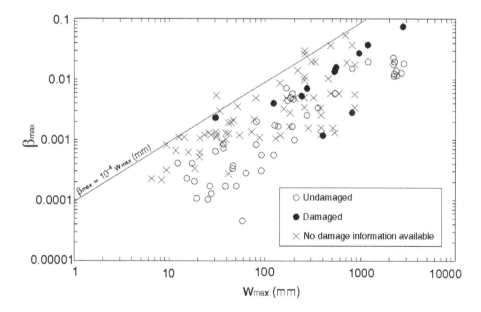

Figure 5.35 Empirical relation between the maximum settlement w_{max} and the maximum expected value of the angular distortion β_{max}, including both shallow and pile foundations.

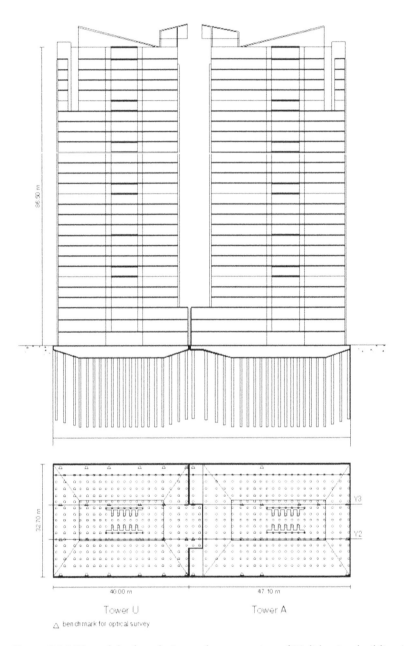

Figure 5.36 Plan of the foundation and cross section of Holiday Inn building in Naples.

spread underneath the rafts (Figure 5.36). The weight of each tower was around 200 MN with the raft accounting for almost half of this value. The superstructure consists of steel frames with reinforced concrete stiffening cores. About 66% of the total load is transmitted via the stiffening cores, while the steel columns uniformly share the remaining 34%.

The prediction of settlement in the hypothesis of flexible foundation is reported in Figure 5.37; it may be seen that the average settlement is correctly predicted, while the differential settlement is substantially overestimated.

Repeating the analysis by the code NAPRA, which will be presented in Chapter 6, taking into account the stiffness of the raft, substantially improves the differential settlement prediction.

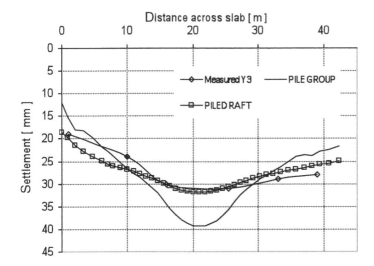

Figure 5.37 Measured and predicted differential settlement.

6   Soil–structure interaction and the design of pile cap

6.1 Introduction

The scope of a soil–structure interaction analysis for a piled foundation is the prediction of the internal forces in the structure connecting the piles (cap or raft) for its structural design, and the evaluation of differential displacements for their possible effects on the superstructure.

To introduce the matter, let us consider the simple case represented in Figure 6.1, in which the pile cap is clear of the soil and the foundation is subjected to a load distribution whose resultant is a vertical force V with eccentricity e_x and e_y from the centre of gravity of the pile group. This problem is routinely analysed assuming that the cap is rigid, neglecting the interaction among piles and assuming that each pile behaves as an independent elastic element (a spring, or an axially loaded pole) exerting on the cap only a vertical reaction.

Due to the supposed rigidity of the cap, irrespective of their actual distribution, the applied loads can be substituted by their resultant, and hence by the sum of an axial load V and two moments $M_x = Ve_x$ and $M_y = Ve_y$.

Under the axial load V the cap undergoes a vertical displacement and hence all the piles have the same settlement; if the piles in the group are equal, the load Q_i acting on each pile will be equal to V/n, being n the number of piles in the group.

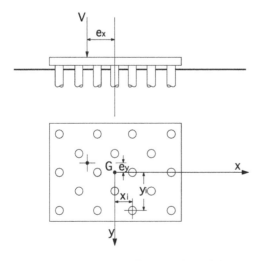

Figure 6.1 Pile group with a cap clear of the ground subjected to an eccentric vertical force.

Under the moment Ve_x, the cap undergoes a rigid rotation α_x around the y axis. The settlement of pile i (x_i, y_i) is equal to $w_i = \alpha_x x_i$; the load $Q_i = kw_i = k\alpha_x x_i$, where k is the axial stiffness of the pile. Equilibrium to rotation around y axis requires:

$$Ve_x = \sum_{i=1}^{n} Q_i x_i = k\alpha_x \sum_{i=1}^{n} x_i^2$$

and hence:

$$\alpha_x = \frac{Ve_x}{k \sum_{i=1}^{n} x_i^2}$$

Substituting into the expression $Q_i = k_i \alpha_x x_i$ gives:

$$Q_i = \frac{Ve_x}{\sum_{i=1}^{n} x_i^2} x_i$$

Superimposing the three load conditions one gets finally:

$$Q_i = \frac{V}{n} + \frac{Ve_x}{\sum_{i=1}^{n} x_i^2} x_i + \frac{Ve_y}{\sum_{i=1}^{n} y_i^2} y_i \tag{6.1}$$

Eq. 6.1 is a discrete version of the expression of the stress in a section acted upon by a normal force and a moment. The model assimilating the piles to independent elastic elements may be regarded as a discrete version of the well-known Winkler model for shallow foundations; it shares with Winkler's the shortcoming of neglecting the mutual interaction among the different elements.

In Chapter 5, in fact, it has been shown that the settlement of a pile does not depend only on the load directly applied to it, but also on those applied to other surrounding piles. Each pile in the group, in fact, interacts with all other piles via the deformation field induced in the soil; neglecting this feature, we predict a behaviour of the pile group in disagreement with the available experimental evidence.

To demonstrate this, let us consider the simple case of a centred load V $(e_x = e_y = 0)$. In this case, Eq. 6.1 reduces to:

$$Q_i = \frac{V}{n}$$

In other words, all the piles in the group would be equally loaded, and the settlement of the group would be equal to the settlement of a single pile subjected to the load V/n.

Both these conclusions are contradicted by the available experimental evidence, since it is well known that the settlement of a pile group increases with increasing the size of the group and that the distribution of the load among the piles of a group with a rigid cap is not uniform, but the corner and peripheral piles are subjected to loads higher than average.

The first effect has been discussed in detail in the previous Chapter 5; it is a consequence of the interaction among the piles via the surrounding soil, as schematically shown in Figure 5.15.

The second effect is illustrated by the experimental data collected in Figure 6.2 (Russo 1998b), showing that the ratio of the load on corner pile to that on centre pile may be as high as four, and that of edge pile to centre pile as high as three, both decreasing with increasing s/d and approaching unity for $s/d \geq 8$.

If the soil is modelled as a continuum (for instance, a linearly elastic half space), both the effects are reproduced by the analysis, at least qualitatively.

The increase of settlement with the size of the group, predicted by modelling the soil as an elastic continuum, has been reported in Figure 5.26.

The load distribution among the piles belonging to groups with a rigid cap, as obtained by an analysis modelling the soil as an elastic continuum, is reported in Figure 6.3. The corner piles are subjected to a load larger than average, while the central piles are acted upon by a load smaller than average and, for small spacing, even tensile. The load on the edge piles is intermediate. With increasing spacing, when the interaction among piles tends to vanish, all loads tend to equalize at the average value. The load concentration on peripheral piles may be explained by referring to Figure 5.15, which shows that in a group composed by equally loaded piles (without a cap, or with an infinitely flexible one) the settlement is not uniform, but larger for the central piles and progressively decreasing for the peripheral piles. It is intuitively clear that when a rigid cap imposes on the piles a constant settlement, the loads on the peripheral piles have to increase and those on the central ones to decrease. More than 30 years ago Burland *et al.* (1977) stated that: "This feature has been observed sufficiently frequently in practice, and has such a strong theoretical basis, that it seems entirely justifiable to take account of it in design."

6.2 Design of the cap of small pile groups

In the simple case of small groups of piles located at the corners of a regular polygon and loaded by a concentrated force at the centre (Figure 6.4), for symmetry reasons the external load subdivides in equal parts on each pile, independently of the subsoil model adopted; the internal forces in the cap may be obtained accordingly by mere statics.

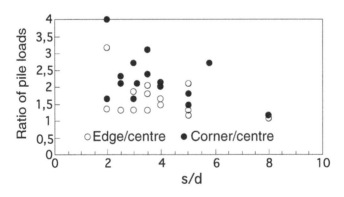

Figure 6.2 Experimental data on the edge effect (data by Russo 1996).

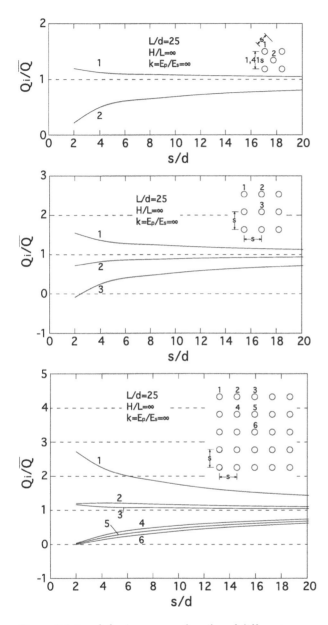

Figure 6.3 Load sharing among the piles of different groups.

The reinforcement in the cap of a two piles group (Figure 6.5) may be determined either in bending or by the so-called method of strut and tie (STM). The former procedure is preferable for $h < s/2$; the section A_f of the tensile reinforcement is obtained by the expression:

$$A_f = \frac{M}{0.9h\sigma_f} = Q\frac{s-a}{1.8h\sigma_f}$$

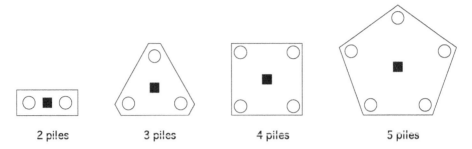

Figure 6.4 Piles located at the corners of regular polygons.

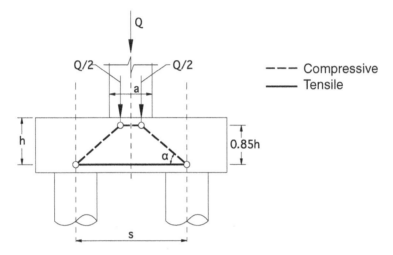

Figure 6.5 Sketch of a plinth on a two piles group.

where σ_f is the allowable tensile stress of the steel. If $h > s/2$, with the STM the reinforcement is designed to resist a tensile force $T = Qt\,g\alpha$; one gets:

$$A_f = \frac{T}{\sigma_f} = \frac{Qtg\alpha}{\sigma_f} = Q\frac{\dfrac{s}{2} - \dfrac{a}{4}}{0.85h\sigma_f} = Q\frac{s - \dfrac{a}{2}}{1.7h\sigma_f}$$

For a three piles group (Figure 6.6a) the reinforcement in bending is equal to:

$$A_f = Q\frac{s - \dfrac{a}{2}}{2.95h\sigma_f}$$

while following the STM (Figure 6.6b), one obtains a radial tensile force:

$$H = 0.38\frac{Q}{h}\left(0.58s - 0.25h\right)$$

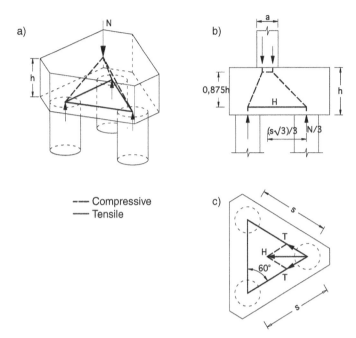

Figure 6.6 Sketch of a plinth on a three piles group.

and hence a radial reinforcement:

$$A_f = Q\frac{0.58s - 0.25a}{2.63h\sigma_f}$$

The radial force H can be decomposed along the perimeter of the cap (Figure 6.6c) obtaining a tensile force parallel to the sides of the cap:

$$T = \frac{H}{2\cos 30°} = 0.22\frac{Q}{h}(0.58s - 0.25h)$$

corresponding to a reinforcement parallel to the sides of the cap:

$$A_f = Q\frac{0.58s - 0.25a}{4.55h\sigma_f}$$

In practice, the layout considered is adopted only for the two, three and four pile groups; for larger groups different configurations are usual, as for instance those reported in Figure 6.7.

In groups of this type, and still more in groups with a larger number of piles, the effect of load concentration on peripheral piles tends to increase the bending moments and shear in the cap. To evaluate the significance of the effect, let us consider some examples.

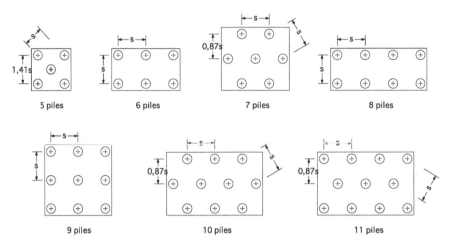

Figure 6.7 Different configurations for small group of piles.

For the sake of simplicity, we will assume that the pile cap is rigid; this allows the determination of the load distribution among the piles independently of the distribution of external loads, as a function only of its resultant and location in plan (Eqs 5.19 and 5.20). Once determined the loads acting on the piles, bending moments and shears may be obtained by statics taking into account the external loads also with their actual distribution.

Let us first consider a group of five piles (Figure 6.8) four of which are placed at the corners of a square and the fifth in the centre, with a spacing of $3d$ axis to axis along the diagonal; the group is loaded by a concentrated force Q at the centre. The simplified analysis of the Winkler type (Eq. 6.1) brings us to the conclusion that each pile is subjected to an axial load equal to $0.2Q$. For the determination of the internal forces in the cap, the load on the central pile may be directly subtracted from the external load; the reinforcement may be evaluated considering a group of four piles

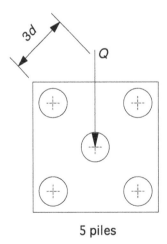

5 piles

Figure 6.8 Scheme of a five piles group loaded by a centred load Q.

subjected to the resulting net load, equal to 0.8Q, equilibrated by the loads on the four external piles, each bearing 0.2Q.

If we model the subsoil as an elastic half space and consider rigid piles with $L/d=25$ (Figure 6.3), we obtain that the load acting on the central pile is equal to 0.07Q, while those on the external piles to 0.23Q. Accordingly, the reinforcement may be evaluated considering a group of four piles subjected to the resulting net load, equal to 0.93Q, with an increase of 16% with respect to the simple Winkler type analysis.

Let us now analyse a square group of $5\times5=25$ piles, again with a rigid cap and a spacing of the piles equal to $3d$ axis to axis. The group is subjected either to a concentrated load Q at the centre or to a uniformly distributed load with constant intensity $q = \dfrac{Q}{(13d)^2}$. The subsoil is modelled either as a half space (HS) or as an elastic layer (EL) of finite thickness $H=1.2L$.

The total moment acting across a line through the cap centre and parallel to the sides is given in Table 6.1.

It may be seen again that the difference between the Winkler type and the elastic continuum models is relatively low, albeit significant, for a concentrated load; it is lower for the less interactive model of elastic layer than for the half space. On the contrary, the difference is enormous for distributed load.

6.3 Design of the raft for large pile groups; the code NAPRA

6.3.1 Introduction

In the case of a large pile group, the bending moments in the raft will be different, and probably somewhat smaller, than those obtained above by assuming the cap rigid and clear of the soil and modelling the subsoil as an elastic continuum, for the influence of the following factors:

- the cap has a finite stiffness, modifying the distribution of load among the piles in the sense of increasing the load on the piles directly below the external load and decreasing that on the peripheral piles;
- the cap or raft is actually not clear of the soil, but resting on it and thus transmitting a portion of the external load directly to the soil;
- non-linearity effects will tend to smooth out the load peaks on the piles (Caputo and Viggiani 1984) and hence to make the bending moments smaller;

Table 6.1 Bending moment in the cap of a square group of $5\times5=25$ piles

Load	Total bending moment			$\dfrac{M_{EL}-M_W}{M_W}$ (%)	$\dfrac{M_{HS}-M_W}{M_W}$ (%)
	M_W (Winkler)	M_{EL} (Elastic Layer, $H=1.2L$)	M_{HS} (Half Space, $H=\infty$)		
Concentrated at the centre	45.0Qd	50.7Qd	57.9Qd	12.7	28.7
Uniformly distributed	4.4Qd	10.1Qd	17.3Qd	129.5	293.2

- in the long term, the pile loads tend to even out because of creep effects in the piles and in the cap (Bowles 1988; Mandolini *et al.* 2005). The cap, however, has to be designed to support the worst loading case even if transient.

In the following articles we will explore the significance of the first three factors, by means of a parametric study carried out by a code named NAPRA (Russo 1996, 1998a), which allows their consideration. It is an extension of Gruppalo to account for a finite flexural rigidity of the raft and direct raft–soil interaction.

The code can solve piled rafts of any shape and flexural stiffness supported by the soil and by piles in any number and pattern, and subjected to any combination of vertical distributed or concentrated loads, and to moment loading. The raft is modelled as a two-dimensional elastic body using thin plate theory, while piles are modelled as non-linear springs mutually interacting through the soil modelled as a linearly elastic body. The interaction between the raft and the soil is lumped into a number of discrete points and it is assumed to be purely vertical.

The lumped stiffness of the soil is deduced by closed form solution for the settlement of a uniformly loaded rectangular area at the surface of a homogeneous elastic half space; the so-called Steinbrenner approximation is used to simulate a horizontally stratified medium. Tension cannot occur at the raft–soil interface; an iterative procedure sets to zero any tensile force developed between raft and soil.

The non-linear load settlement relationship of the piles is simulated by a stepwise incremental procedure.

When used to predict settlement of pile foundations designed according to the capacity based approach, NAPRA gives essentially the same results as Gruppalo; in addition, it provides the load sharing between piles and raft, the load distribution among piles and the bending moment and shear force in the raft.

6.3.2　FEM analysis of the raft

The raft is analysed in NAPRA by the Finite Element Method, adopting a four node rectangular element (Griffith *et al.* 1991; Zienkiewicz and Taylor 1991). The element is based on the thin plate theory, which does not allow for transverse shear strain. Any geometry of the raft can be modelled by rectangular elements adopting a piecewise approximation.

The equation of bending for thin plates may be written as follows:

$$D\nabla^4 w(x,y) = q(x,y)$$

where $w(x,y)$ is the unknown vertical displacement of the raft, $q(x,y)$ is the applied load and $D = \dfrac{E_r t^3}{12(1-v_r^2)}$ the flexural stiffness. In the expression of D, E_r and v_r represent the Young's modulus and Poisson's ratio of the raft, t its thickness.

In the Finite Element approach, the above equation is written in terms of a finite number of nodal displacements as follows:

$$[K_r]\{w_r\} = \{q\} \tag{6.2}$$

where $[K_r]$ and $\{w_r\}$ are respectively the stiffness matrix and the vector of nodal displacements of the raft, and $\{q\}$ represents the vector of nodal forces or moments acting on the raft.

As the four node rectangular element used in the analysis has four degrees of freedom at each node, the stiffness matrix of the raft is a square $4n \times 4n$ matrix, where n is the number of nodes used to model the raft. If any boundary of the raft is constrained, the stiffness matrix incorporates the related boundary conditions.

6.3.3 Closed form solution for soil displacements

The soil supporting the raft is modelled as an elastic continuum; the soil displacements produced by the contact pressure developed at the interface between the raft and the soil are obtained by Boussinesq (1885) solution for a point load and the closed form solution for a rectangular uniformly loaded area (Harr 1966) at the surface of a homogeneous elastic half space. The Boussinesq solution is used to calculate the displacement w_{ij} occurring at a point J due to the contact pressure developed in the i-th element, whose resultant is lumped in its central point I. The displacement w_{ii} occurring at the point I due to the pressure acting on the same element i is obtained by the solution for the rectangular uniformly loaded area.

The horizontally layered continuum is solved by a repeated application of Steinbrenner's approximation, which basically assumes that the stress distribution within an elastic layer bounded by a rigid base is identical with that occurring in a homogeneous half space. The accuracy of Steinbrenner's approximation has been explored by different authors (Davis and Taylor 1961; Poulos and Davis 1980; de Sanctis *et al.* 2002), and shown to be generally acceptable for engineering purposes.

The flexibility matrix $[F_s]$ of the soil is built up by the above solutions; it relates the vector $\{w_s\}$ of the unknown nodal soil displacements to the nodal vertical interaction forces $\{r_{rs}\}$ as follows:

$$[F_s]\{r_{rs}\} = \{w_s\}$$

6.3.4 Piles as non-linear interacting springs

A considerable computational simplification for the analysis of piled rafts is obtained if each pile is considered as a single unit whose reaction is lumped into a node of the thin plate. According to Caputo and Viggiani (1984), the non-linearity of the pile–soil interaction overwhelms all the other non-linearity factors of the response of a piled raft; the same view is expressed by Randolph (1994) and El Mossallamy and Franke (1997). In the code NAPRA, the non-linear load–displacement response of the single pile is modelled by an hyperbola (Chin 1970):

$$Q = \frac{w}{m + nw}$$

A broad experimental evidence shows that most of the available load test on piles can be closely fitted by this expression in which $1/n$ is equivalent to the ultimate bearing capacity of the pile and $1/m$ to the initial tangent modulus of the load–settlement curve.

The interaction among piles through the elastic continuum is modelled by the method of interaction factor, as already described in Chapter 5.

6.3.5 Interaction between piles and raft elements

The interaction between an axially loaded pile beneath the raft and an element of the raft develops through the elastic continuum. A BEM procedure is implemented in NAPRA to calculate a pile–soil interaction factor α_{ps} defined as follows:

$$\alpha_{ps}(s) = \frac{w_2(s)}{w_p}$$

where $w_2(s)$ represents the displacement at the centre of a raft element located at a distance s from the pile and w_p is the displacement of the pile head. The interaction factors are computed at various spacing and fitted with a continuous curve of equation:

$$\alpha_{ps}(s) = \frac{1}{1 + A\left(\dfrac{s}{d}\right)^B} + C\log\left(\frac{s}{d} + 10\right)$$

in which the unknown parameters A, B and C are determined by fitting.

Some authors (Hain and Lee 1978; Poulos 1994) have assumed that the pile–soil interaction factors are equal to the pile–pile ones; it can be easily shown that this approximation can be very rough. In the code NAPRA the reciprocal theorem is used only to assume that the soil–pile interaction factor is equal to the corresponding pile–soil one.

6.3.6 Solution procedure

The assumption that the mutual interaction between the raft and the pile–soil system is purely vertical reduces the size of the stiffness matrix of the raft to $n \times n$ by means of a partial backward substitution. Eq. 6.2, for the raft subjected to the external loads and to the pile–soil reactions $\{r_{sr}\}$, may be written as follows:

$$[K_r]\{w_r\} = \{q\} + \{r_{sr}\} \tag{6.3}$$

where the same symbols of Eq. 6.2 have been used, even if both the stiffness matrix and the external load vector have been reduced to the vertical degrees of freedom only.

The stiffness matrix $[K_{sp}]$ of the pile–soil system is obtained by inversion of the flexibility matrix $[F_{sp}]$. If the pile–soil system is subjected to the raft nodal reaction $\{r_{rs}\}$ then:

$$\left[K_{sp}\right]\{w_{sp}\} = \{r_{rs}\} \tag{6.4}$$

The compatibility of the displacements of the raft and the pile–soil system is expressed by the following relationship:

$$\{w_r\} = \{w_{sp}\} = \{w\}$$

and the addition of Eqs 6.3 and 6.4 yields:

$$[K]\{w\} = \{q\}$$ (6.5)

where $[K] = [K_r] + [K_{sp}]$.

The linear system (Eq. 6.5) is then solved for the unknown displacements. Because of the non-linear load–settlement relationship used for the piles, a stepwise linear incremental procedure is implemented in the program; it subdivides the total load to be applied into a number of increments, and the diagonal terms of the pile–soil flexibility matrix are updated at each step, according to the equation reported in §5.3.5.

The nodal reactions vector $\{r_{rs}\}$ is computed at each step, to check for tensile forces between raft and soil, and an iterative procedure is used to make them equal to zero. Basically this procedure releases the compatibility of displacements between the raft and the pile–soil system in the node where tensile forces were detected, although the overall equilibrium is saved by means of a partial backward substitution. An iterative procedure is needed since, after the first run, some additional tensile forces may arise in different nodes.

The output of the code is represented by the distribution of the nodal displacements of the raft and the pile–soil system, the load sharing among the piles in the group and the soil, and the bending moments and shear in the raft, for each load increment.

6.4 Influence of the finite stiffness of the cap

To explore the influence of a finite stiffness of the cap, the piled rafts reported in Figure 6.3 have been analysed by the code NAPRA considering a cap of finite stiffness. Values of the relative raft–soil stiffness K, as defined by Wardle and Fraser (1974):

$$K = \frac{4E_r t_r^3 (1 - v_s^2)}{3E_s B_r^3 (1 - v_r^2)}$$

ranging between 10^{-4} and 10^4 have been considered. The load sharing among the piles belonging to the different groups are plotted in Figure 6.9 for the case of uniform applied load and in Figure 6.10 for the case of a concentrated axial force; of course, the consideration of a cap with a finite stiffness makes the results depending on the type of load distribution. As expected, when the value of K increases the solutions reported in Figures 6.9 and 6.10 tends to coincide (load sharing does not depend on the distribution of the applied load); furthermore, the results tend to the solutions obtained via the code Gruppalo which assumes infinitely stiff cap.

Both figures show how large the influence of the relative raft–soil stiffness may be on the computed load sharing. Moving towards lower K values produces initially a substantial smoothing out of the peak load on the peripheral piles. For uniform load (Figure 6.9), K values around 1 usually determines equal loads on all the piles; in the case of concentrated central force (Figure 6.10), this occurs at K values between 5 and 10.

In the left part of the plots, for the lowest investigated values of K and hence for an infinitely flexible raft, the load on the *i-th* pile approaches the load directly applied on the tributary of the pile.

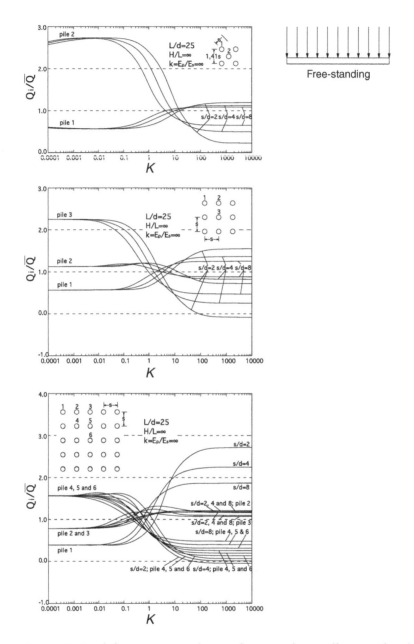

Figure 6.9 Load sharing among piles as a function of cap stiffness: uniformly distributed loading.

All the above remarks apply qualitatively to the three investigated layouts, corresponding to different pile–pile spacing (s/d = 2; 4; 8). The larger the spacing, the lower the interaction among the piles, with the obvious consequence of reducing the load concentration on peripheral piles (edge effect) for the cases with higher values of relative raft–soil stiffness K.

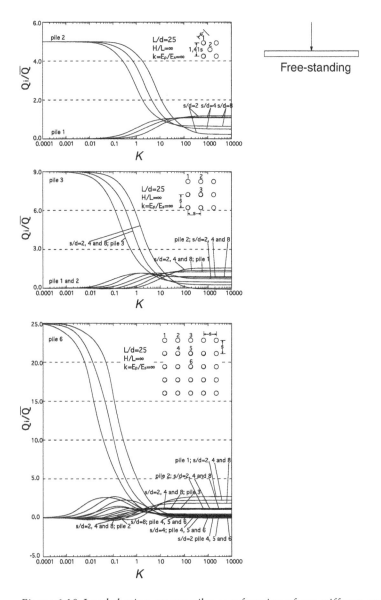

Figure 6.10 Load sharing among piles as a function of cap stiffness: concentrated central force.

The maximum bending moments at the centre of the raft are plotted in Figures 6.11 and 6.12, respectively for uniform applied load and concentrated central force; bending moments are assumed positive when the lower edge of the raft is subjected to tensile stress.

The relative raft–soil stiffness has a large influence on the computed bending moment also. Similarly to the findings for unpiled rafts, the higher the stiffness, the higher the bending moments; for the same relative raft–soil stiffness and for the same

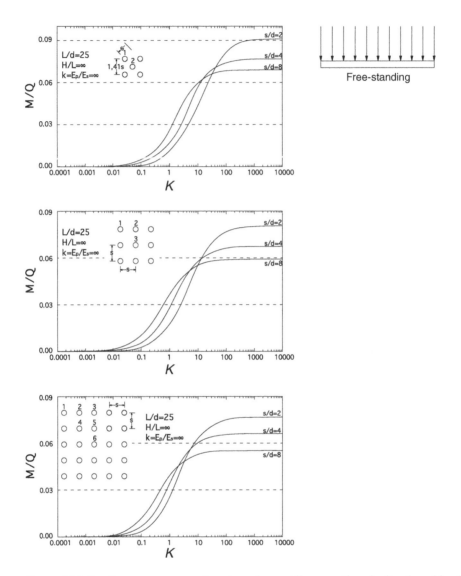

Figure 6.11 Bending moments as a function of cap stiffness: uniformly distributed load.

pile group the higher the spacing (i.e. the higher the stiffness of the pile group), the lower the bending moment in both cases of uniform load and single concentrated central force.

For the uniform load case (Figure 6.11) at K values around one the bending moment at the raft centre is significantly lower than the value for infinitely stiff rafts, and approaches zero. Such a finding corresponds to the uniform load sharing among the piles of the group shown in Figure 6.9. The same does not occur for the concentrated load; in this case only for an infinitely flexible raft, when the load is absorbed by the only central pile, the maximum bending moment approaches indeed to zero.

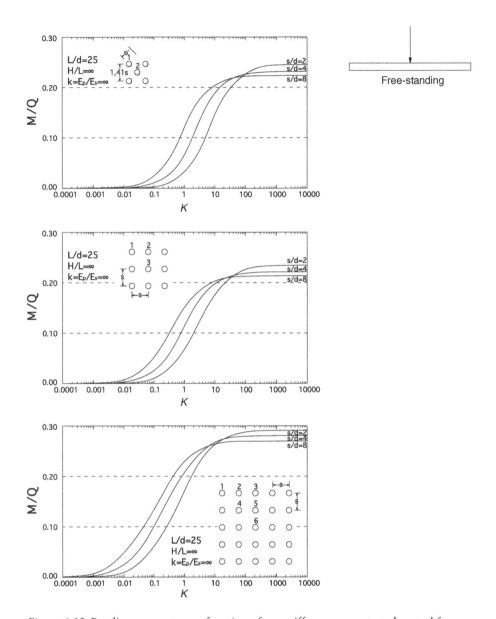

Figure 6.12 Bending moments as a function of cap stiffness: concentrated central force.

6.5 Influence of cap in contact and non-linearity of the piles

The cap of a piled raft is actually in contact with the soil and thus capable of transmitting part of the external applied load directly into the soil. The stress at the interface between the raft and the soil influences the internal forces in the raft, modifying the values computed in the assumption of a raft clear of the soil and resting only on the piles.

In Figures 6.13 and 6.14 the load sharing among the piles and the maximum central bending moments in the raft are plotted for the same cases analysed in §6.4, but taking into account the raft–soil contact. For sake of simplicity, only the case of uniform load has been considered.

Comparing Figure 6.13 to Figure 6.9, it appears that the load sharing among the piles is nearly unaffected by the cap being in contact, at least for values of $K \geq 1$. The soil does not significantly participate in the global load sharing with the cap; for the cases $s/d = 2$ and $s/d = 4$ most of the load is taken by the piles. At K values between

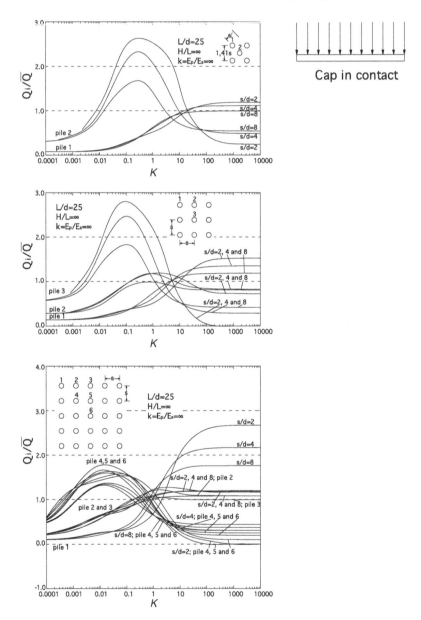

Figure 6.13 Load sharing among piles below a cap in contact with the soil.

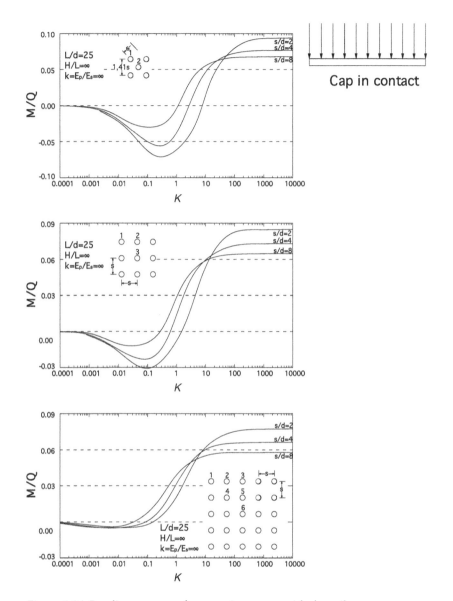

Figure 6.14 Bending moments for a cap in contact with the soil.

0.1 and 1 the load sharing approaches the same values obtained without the cap in contact, which correspond to the loads computed on the basis of tributary area. Only when $s/d = 8$ there is a significant participation of the soil to the global load sharing.

For the lowest investigated values of K, approaching an infinitely flexible raft, in all the cases there is a significant participation of the soil contact to the global load sharing; the load on piles is approximately equal to a fraction of the external applied load corresponding to the percentage of the area of the raft directly occupied by the piles.

In terms of bending moments, the positive values calculated for stiff rafts are in all cases similar but a bit larger than those computed with the cap clear of the soil. Differently from the case of the cap clear of the soil, for values of K between 0.1 and 1 the bending moments in the central section become negative (i.e. the upper edge of the raft is subjected to tensile stress), the negative values being always smaller than the positive values obtained for stiff rafts. Only for K values around 0.001 (infinitely flexible raft) the bending moments vanishes.

The non-linearity of the piles is another relevant factor influencing the load sharing and, consequently, the bending moments. Any form of non-linearity tend to smooth out the edge effects on stiff rafts, thus reducing the maximum positive bending moment; some results obtained by Caputo and Viggiani (1984) are reported in Figure 6.15. The loads taken by the corner pile and by the central piles in a 5×5 square group are plotted in this figure as a function of the safety factor SF_g on the pile group. As SF_g approaches unity, the edge effect vanishes.

6.6 Influence of creep

The influence of creep of the piles and/or of the reinforced concrete of the pile cap may be another significant issue in determining, at least in the long term, a substantial

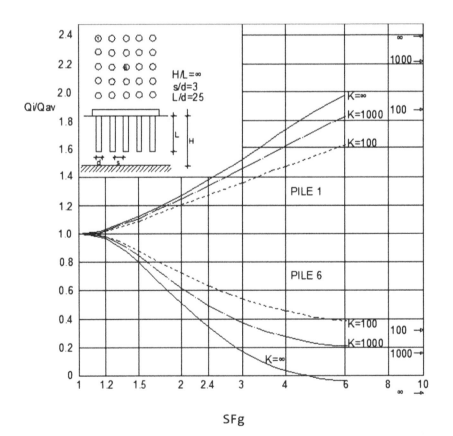

Figure 6.15 Non-linearity effects on load sharing (after Caputo and Viggiani 1984).

smoothing out of the initial edge effect thus reducing the maximum bending moments.

To support this statement the data from a case history are briefly introduced.

The case history is that of the main pier of the cable-stayed bridge across the River Garigliano (southern Italy). The foundation of the main pier, resting on driven tubular steel piles, is represented in Figure 6.16. It was monitored during the construction and afterwards, measuring settlement, load–sharing between piles and raft and load distribution among the piles.

The construction of the bridge started in October 1991 and the latest set of data has been recorded in October 2004, almost 13 years later. Of the 144 piles, 35 were equipped with load cells at the top, to measure the load transmitted by the cap to

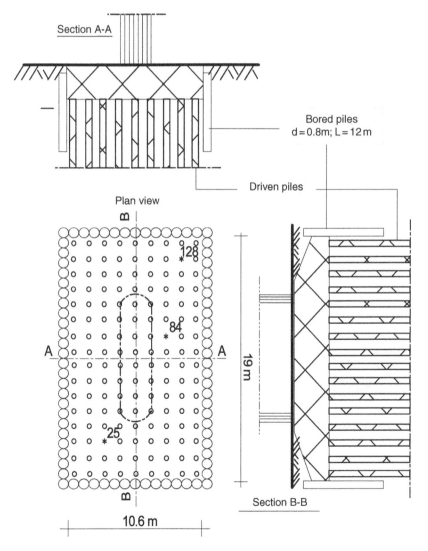

Figure 6.16 Layout of the foundation of the main pier of the bridge over River Garigliano.

the pile; furthermore, eight pressure cells were installed at the interface between the cap and the soil. The load cells and pressure cells were constructed on site using three sensing units for each of them; a total of 129 vibrating wire load sensing units were used. Further details on the instruments and the installation technique are reported by Mandolini and Viggiani (1992) and Russo and Viggiani (1995) (Figure 6.17).

At the end of construction the measurements show a significant edge effect, as was to be expected under a stiff cap, and some load concentration below the pier. Three years later the load distribution had undergone a significant variation: the load on the peripheral piles was decreasing, while that on the piles below the pier was slightly increasing. Ten years later, this trend is confirmed by the values reported in Table 6.2. To the writers' knowledge, such a phenomenon had not been observed before; the observed trend of variation suggests that the main factor is creep of the reinforced concrete raft.

In Figure 6.18 the bending moments along three different sections of the pile cap computed by the code Gruppalo are compared with those computed by the code

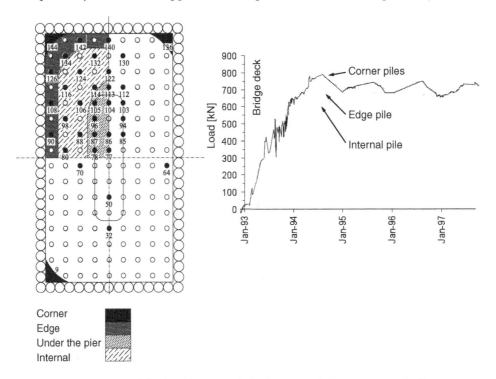

Figure 6.17 Plan view of the foundation with the location of the instrumented piles.

Table 6.2 Garigliano: load sharing among the piles vs. the time

	Corner piles	Edge piles	Internal piles	Piles under the pier
End of construction	1.30	1.00	0.80	0.90
3 years later	1.16	0.96	0.90	0.98
10 years later	1.10	0.93	0.94	1.03

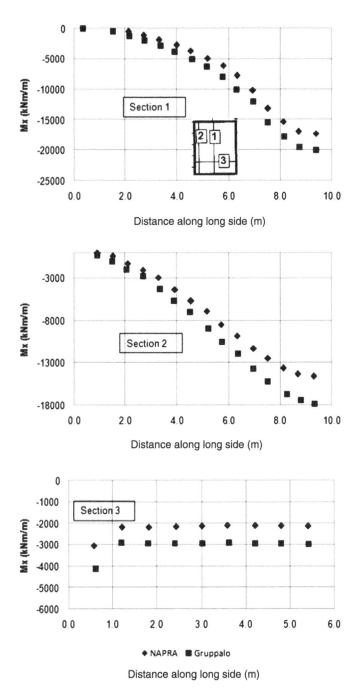

Figure 6.18 Bending moments computed for Garigliano pile cap (after Russo 1997).

NAPRA. The main difference between the two codes is that NAPRA considers the raft–soil contact while Gruppalo assumes a raft clear of the soil. The figure shows that the differences between the two approaches is not that large; the more realistic model implemented in NAPRA, however, brings to moments lower than those computed by Gruppalo, the difference being at worst about 25%. A more detailed discussion on this case history will be reported in Chapter 10.

6.7 Concluding remarks

The design of the pile cap is traditionally based on very simple but rough models of the pile group. It has been shown that for small pile groups (usually subjected to concentrated loads) the simple and still widespread Winkler's approach is not grossly in error. With the usual safety factors implied in structural design, it can be concluded that the routine approach based on the Winkler-type model is fully justified, at least for small pile groups. Such an approach is even exact for caps with piles at the corners of a regular polygon.

For large pile groups, where the cap is usually represented by a raft, the interaction among the piles cannot be neglected. It is responsible for a significant edge effect in the load sharing among the piles, which usually brings to maximum values of the bending moments significantly larger than those obtained by the simple model. The code Gruppalo has been used to show such an effect for rigid rafts clear of the soil.

The code NAPRA takes into account other factors, such as the finite stiffness of the cap and the possibility of the cap to transmit directly load onto the soil; a few examples have been illustrated, just to allow the reader to grasp their influence on the computed internal forces in the cap. It has to be emphasized, however, that every single foundation has its own peculiarity in terms of external loading and/or subsoil condition; the variety of possible situations is such that no generalization is possible. The availability of suitable codes (as for instance NAPRA) allows the designer to run a proper analysis in each particular case.

7   Pile testing

7.1 Vertical load tests

7.1.1 General

According to their purpose, load tests on foundation piles can be subdivided into *design load tests* and *proof load tests*.

The design test is usually kept to failure, or at least to a maximum load not less than three times the intended service load. It is a destructive test, and has to be carried out on a purposely installed test pile, which does not belong to the foundation. The aim of a design load test is to determine, at the design stage, the bearing capacity of the pile and its load–settlement relationship; if the pile shaft is properly instrumented, it allows the determination of the transfer curves of the side shear (the so-called τ–z curves) and base load (p–z curve), and the fractions of the bearing capacity taken by the base and the shaft of the pile.

The proof load test, on the contrary, is carried out on piles selected among the piles of the foundation, after they have been all installed. The test cannot be destructive, and hence the maximum test load is usually limited to 1.5 times the intended service load. The number of piles to be proof tested is generally specified in the tender documents; in Italy it ranges usually between 1% and 2% of the total number of piles, with a minimum of two. It is aimed at verifying the correct installation of the piles, but indications on the load–settlement behaviour and, by extrapolation, on the bearing capacity may be also obtained.

In the case of design tests, the installation of the test pile must reproduce as closely as possible that of the production piles in order to obtain representative results; if possible, the same specialized contractor should install the test pile and the production piles. For the same reason the test pile has to be installed as close as possible to the location of the foundation. A borehole or a CPT is carried out in the vicinity of the test pile, in order to know the exact subsoil profile at the test site.

The piles to be proof tested are selected only after all the piles have been installed, in order to prevent a particularly careful installation of the intended test pile and to obtain an equal care for all the piles.

Driven piles in cohesive soils have to be tested after the excess pore pressure induced by driving has substantially dissipated; this involves a delay of some days to some weeks after driving.

7.1.2 Test equipment

The test equipment and procedures are essentially the same for design and proof load tests. The load is applied by means of a hydraulic jack, resting on the pile head; to this aim the pile head must be plane and horizontal. For cast *in situ* piles, additional hoop reinforcement or a steel tube casing about two diameters long is provided at the pile head.

The reaction to the jack is provided either by a kentledge or by a reaction beam anchored to the ground by tension piles or ground anchors. The kentledge rests on two lateral support walls through a bed of steel beams (Figure 7.1); in order not to influence the behaviour of the test pile, the spacing between the supports and the test pile should be not less than 2 m or two times the pile diameter. The total weight of the kentledge should be 10% more than the maximum test load, and carefully centred over the test pile. If an anchored beam is used as reaction system (Figure 7.2), the tension piles should also be sufficiently spaced; further indications are provided in §7.1.4.

The hydraulic jack is pressurized by an oil pump; the load can be kept constant either by continuous manual regulations or by providing the pump with an automatic adjustment controlled by the load measurement.

In principle, the applied load could be obtained by the oil pressure multiplied by the area of the jack ram; such a measurement, unfortunately still in use, is very rough due to the inevitable and somewhat random friction in the jack. A correct practice requires the use of a load cell, interposed between the jack piston and the reaction system.

The settlement of the head of the test pile is measured by dial gauges or displacement transducers (or both), fixed to the pile head and contrasting on one or two reference beams, anchored to supports sufficiently spaced from the test pile (Figure 7.3). It is recommended that redundant settlement reading be taken during the test by optical levelling referred to a benchmark located in the vicinity of the test pile.

Figure 7.1 Load test setup with kentledge.

Figure 7.2 Load test setup with tension piles and spreader beam.

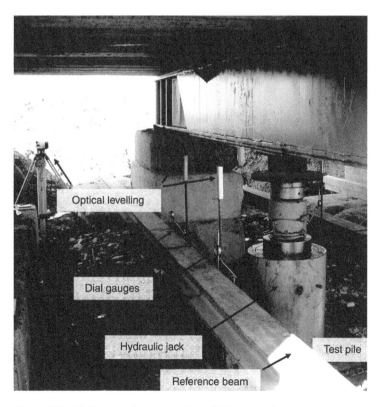

Figure 7.3 Dial gauges for measuring pile head settlement.

Dial gauges or displacement transducers allow settlement to be read to the approximation of 0.01 mm or better, but are influenced by the possible movement of the reference beams and their support and by temperature variations; optical levelling guarantees an approximation of 0.1 mm but is more stable and reliable.

The results of a typical load test are reported in Figure 7.4.

The amount and value of information that can be obtained by a pile load test increase substantially if the displacements or the strain along the pile shaft are measured, in addition to the pile head settlement. The vertical displacement of a number of points at various depth in the pile shaft may be measured by tell tales (Figure 7.5) or by multipoint extensometers (Figure 7.6). The strain may be determined by strain

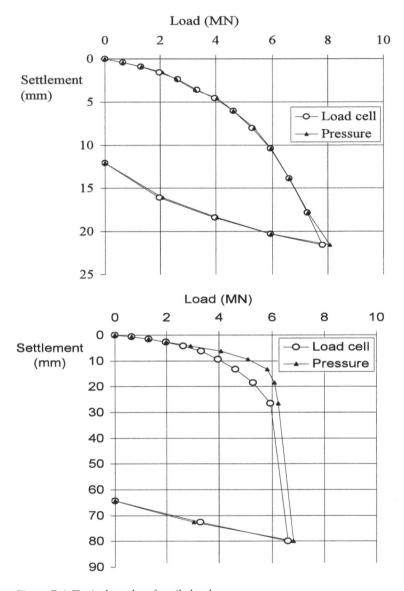

Figure 7.4 Typical results of a pile load test.

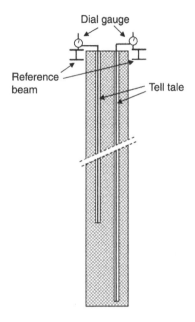

Figure 7.5 Measurements of the displacement of points at depth in the pile shaft by tell tales.

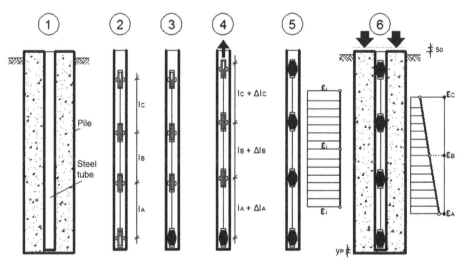

1. Pile ready to be tested; **2.** The m.p.e. is introduced in the tube; **3.** The bottom anchorage is expanded and blocked; **4.** The copper-berillium ribbon is tensioned; **5.** All the other anchorages are blocked; **6.** Load test

Figure 7.6 Measurements of the displacement of points at depth in the pile shaft by multi point extensometer (after Bustamante *et al.* 2003).

gauges fitted to the reinforcement cage (Figure 7.7). The kind of results that are obtained are exemplified in Figure 7.8.

7.1.3 Test procedure

The most common test procedure is the so-called maintained load test, in which the load is applied in steps and each load step is kept constant for a certain time while the settlement is measured. Each load increment should be around 25% of the

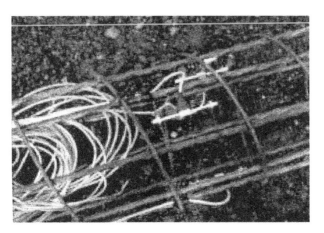

Figure 7.7 Strain gauges fitted to the reinforcing cage.

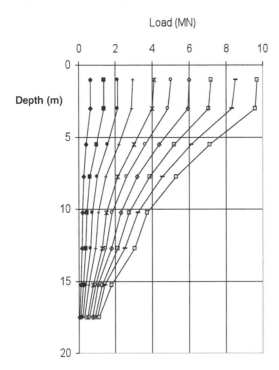

Figure 7.8 Typical results of a load test on an instrumented pile.

intended design load or smaller; so doing, at least six increments are obtained in a proof test and 12 in a design test, which allow a good definition of the load–settlement relationship. The duration of each increment may be fixed prescribing that the next load step may be applied when the settlement has attained its final value. In practice, it is customary to fix a conventional criterion; for instance, it is assumed that the settlement has reached its final value when two readings, taken at an interval of 20 minutes, do not differ more than 0.01 to 0.03 mm. This procedure makes impossible to evaluate in advance the duration of the test; this shortcoming may be avoided by adopting a fixed duration for each load step. A duration of the order of one hour is convenient, and produces a load–settlement curve hardly distinguishable from that obtained by the previous procedure.

A longer duration of the load increment corresponding to the service load is often prescribed. This practice does not provide significant information on the settlement-time relationship of the foundation, that depends on the possible occurrence of deep compressible layers, and produces an irregularity of the load–settlement curve that makes its interpretation more difficult; accordingly, it should be abandoned. The practice of performing intermediate unloading-reloading cycles is also of doubtful usefulness.

It is suggested to adopt a load history consisting in a single loading cycle from zero to the maximum test load, followed by unloading to zero. In the unloading stage, unloading steps larger and duration shorter than those in the loading stage may be adopted.

Slightly different procedures are suggested by a number of authorities such as ASTM (D1143 81) or ISO (ISO/DIS 22477–1)

7.1.4 Test interpretation

Load–settlement relationship

Until a few years ago the aim of a load test was essentially the determination of the bearing capacity, to be employed in a capacity based design. Recently, attention has been switched to the settlement prediction and, accordingly, the scope of pile load tests has broadened to include the determination of the whole load–settlement relationship.

The static vertical load test is generally confused with the Ideal Load Test (ILT) (Figure 7.9a). As we have seen in §7.1.2 above, in practice the load is applied to the pile by a hydraulic jack; the reaction system can be a kentledge resting on supports (Figure 7.9b) or a beam anchored to the soil by tension piles or ground anchors (Figure 7.9c). Recently the so-called Osterberg cell (Figure 7.9d), providing a "self-reaction", is becoming increasingly popular. The setups illustrated in Figures 7.9b–7.9d differ from the ILT because they apply to the ground a load system with zero resultant. This is not without consequences on the load–settlement relationship and on the ultimate bearing capacity, as compared to that of ILT.

In the case of a test with kentledge Poulos (2000) notes that the stress arising in the subsoil from the weight of the kentledge tends to cause an increase of the shaft friction and end bearing of the pile. As the load on the pile is increased by jacking against the kentledge, the stress will reduce and some upward displacements tend to develop in the soil, while the pile undergoes settlement. The pile head stiffness is thus

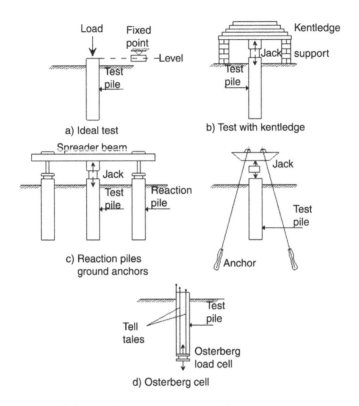

Figure 7.9 Various load tests setup and ideal test.

overestimated, while the pile's ultimate capacity may be relatively close to that of the ideal test.

Figure 7.10 reports the results of a parametric study on the load–settlement curves of a pile subjected to an Ideal Load Test and to a test with kentledge. The curves have been obtained by non-linear finite element elasto-plastic analyses assuming the soil to be a uniform sand with different values of the friction angle φ' (Mohr-Coulomb model) and test piles with $d=1$ m and $L/d=10$, 20 and 50. The test with kentledge overestimates the initial stiffness of the pile, the more the higher the ratio L/d.

On the contrary, at relatively large displacements ($w=10\%\,d$) the discrepancies decrease and eventually the value of the ultimate capacity is practically unaffected by the influence of the kentledge. Similar trends have been found for undrained clays.

The effect of interaction between reaction piles and the test pile is again an over-estimation of the pile head stiffness. The overestimation may be very significant for slender piles and reaction piles close to the test pile (Poulos and Davis 1980; Poulos 2000; Kitiyodom *et al.* 2004). Some further results are reported in Figure 7.11, which refers to a pile with $d=1$ m, $L/d=20$, two reaction piles identical to the test pile and different values of the spacing s between the test pile and the reaction piles ($s/d=4$, 6, 10). Figure 7.11 is based on the same soil model adopted for the analyses reported in Figure 7.10; similar trends have been found for undrained clays.

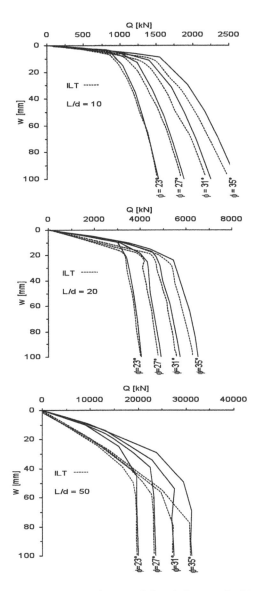

Figure 7.10 Load test with kentledge vs. Ideal Load Test.

In the case of a test pile jacked against ground anchors, Poulos (2000) has shown that the overestimation of the pile head stiffness is significantly less than when reaction piles are used, especially if the anchors are located well below the base of the test pile.

Summing up, the test setups with kentledge or anchor piles are totally suitable for the determination of the bearing capacity. If, on the contrary, the main purpose of the test is the determination of the load–settlement behaviour, and especially the initial stiffness, substantial corrections are needed in all cases. Without these corrections, any analysis based on the load test on single pile can be misleading and unconservative.

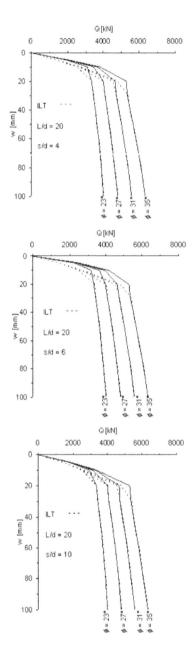

Figure 7.11 Load test with tension piles vs. Ideal Load Test.

Bearing capacity

As reported in §4.2, it is generally accepted that the vertical bearing capacity of a pile is conventionally defined as the load corresponding to a settlement equal to $0.1d$. If the load test has attained such a settlement, the bearing capacity is determined directly in an unambiguous way; for practical reasons, however, such a large settlement is not always attained. In this case, the simplest way of determining the bearing capacity, as defined above, is that of extrapolating the load–settlement curve to a settlement $w = 0.1d$.

Chin (1970) noted that in most cases the load–settlement curve is well approximated by a hyperbola:

$$Q = \frac{w}{m + nw} \tag{7.1}$$

where w is the settlement under the load Q and m, n are two constant to be determined by fitting the curve to the experimental results. Eq. (7.1) may be written:

$$\frac{w}{Q} = m + nw \tag{7.2}$$

suggesting plotting the experimental result in a graph $(w/Q, w)$ and to interpolate a straight line among them (Figure 7.12). The intercept on the vertical axis is equal to m, while the inclination over horizontal is equal to m. Once m and n have been determined, Eq. 7.1 may be used to calculate the bearing capacity:

$$Q_{\text{lim}} = \frac{0.1d}{m + 0.1nd}$$

If the test has been kept to a sufficiently high maximum test load, the extrapolation is very reliable.

The same technique may be employed to extrapolate a proof load test to estimate the bearing capacity. In this case, the maximum load rarely exceeds 50% of the bearing capacity and the results from the extrapolation are much less reliable, but generally conservative.

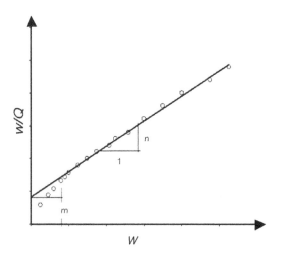

Figure 7.12 The procedure suggested by Chin (1970) to interpolate a hyperbola through the experimental results of a load test.

Transfer curves

Transfer curves of the side shear (τ–z) and base resistance (p–z) can be obtained by interpreting the results of load tests on piles instrumented to measure the displacements or the strain along the shaft. The procedure is illustrated in Figure 7.13.

7.2 Osterberg cell test

The Osterberg cell, commonly called the "O-cell", is a hydraulically driven, high-capacity sacrificial jack-like device composed of two steel plates, with a diameter 50 to 100 mm smaller than the pile diameter, and one or more hydraulic jacks in between (Figure 7.14). The cell is placed in the hole of a cast *in situ* bored pile, in

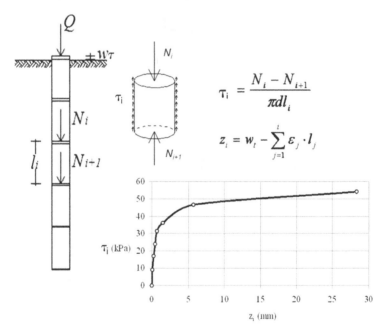

$$\tau_i = \frac{N_i - N_{i+1}}{\pi d l_i}$$

$$z_i = w_t - \sum_{j=1}^{i} \varepsilon_j \cdot l_j$$

Figure 7.13 Determination of the transfer curves τ,z from the results of a load test on instrumented pile.

Figure 7.14 The Osterberg cell.

the vicinity of the base, after having poured in the hole a small amount of concrete to provide a firm setting to the cell; the pile shaft is then concreted.

After the concrete has hardened, the jack in the cell is pressurized in steps. The end bearing support provides reaction for the side friction along the shaft and vice versa, until reaching the capacity of either of the two support components.

During an O-cell test the relative displacement of the two plates is measured by displacement transducers connected to a data acquisition system. Tell tales are used to measure both the compression of the test pile shaft and the upward movement of the top of the O-cell. The downward movement of the bottom plate is obtained by subtracting the upward movement of the top of the cell from the total extension of the cell as determined by displacement transducers. The upward movement of the pile shaft, if any, is measured by dial gauges mounted on a reference beam or by optical levelling.

Typical results of an O-cell test (OCT) are reported in Figure 7.15.

The test goes on until either the base or the shaft reach the ultimate resistance. This is a disadvantage of OCT, since only a lower bound of the total bearing capacity may be determined; the full value is approached only when the two resistances have nearly the same value. The introduction of more than one O-cell at different depths mitigates this limitation, allowing the engineer to obtain both the ultimate end bearing and the ultimate side shear values in cases where the end bearing is less than the side shear (Figure 7.16). On the other hand in the OCT a reaction system is not needed; for this reason it may be a cheap alternative to tests with kentledge or reaction piles and anchors, especially for the high load test needed for large diameter piles.

Osterberg (1995) and Schmertmann and Hayes (1997) give suggestions to derive from the results of OCT a load–settlement curve of the pile head, equivalent to that obtained by the ILT. The suggested procedure relies upon two hypotheses:

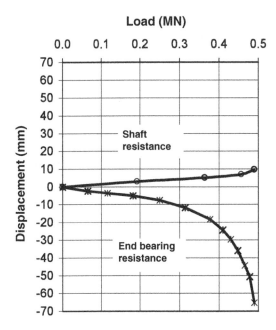

Figure 7.15 Typical results of an O-cell test.

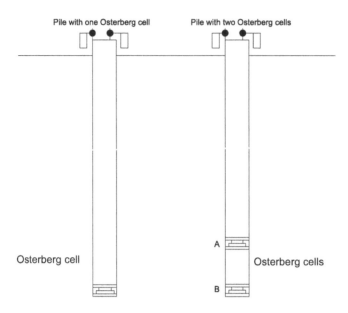

Figure 7.16 Load test with two O-cells along the pile shaft.

- the pile is rigid
- the load–displacement relationship for the shaft resistance is independent of the direction of the relative movement between the pile and the surrounding soil.

A third implicit assumption, which is often neglected, is that the stress and strain fields at the pile base and along the pile shaft are independent of each other and the load–settlement relationship of the base and the shaft can be considered separately.

A parametric FEM analysis has been carried out on this topic (Recinto 2004). The subsoil was assumed as a purely frictional or a purely cohesive, elastic perfectly plastic material. A comparison between the results of a ILT and an O-cell test is reported in Figure 7.17.

A substantial overestimation of the pile head stiffness is again evident at low displacements, while a better agreement is found in the late stage of the test. The shortcoming related to OCT is evident for piles with $L/d = 50$: the end bearing capacity is many times larger than the shaft capacity, preventing the test from exploring the behaviour of the pile further than a settlement $w = 1.5\%d$.

7.3 Dynamic load test

7.3.1 Introduction

Test methods of piles based on the application of some form of dynamic load have been increasingly used in the last decades; among them, the *integrity* test, the *Statnamic* or *rapid* test and the *dynamic* test. Compared to the traditional static load test discussed in the previous paragraphs, they are substantially cheaper and easier to organize.

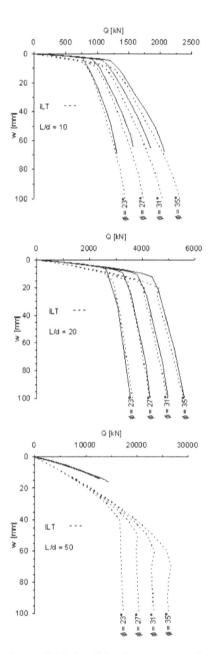

Figure 7.17 O-cell load test compared to Ideal Load Test.

Dynamic test methods are usually classified according to the amount of energy supplied to or the strain level induced in the pile during the test, and have different aims and capabilities in relation to these quantities.

The integrity test keeps in the range of low strain, and is intended only to check the structural integrity of the pile shaft. As such, its application is generally limited to replacement piles, where there is the risk of structural defects connected to the installation procedure.

On the other hand, high strain tests are aimed at determining the bearing capacity of the pile, and secondarily its load–displacement response. Both the dynamic load test, usually made by high energy piling hammers, and the rapid test, widely known as the Statnamic load test, belong to this latter category.

Some features of the various test methods are summarized in Table 7.1. The cost of the test is not reported because of the difficulty of expressing it with a single figure, but indeed it decreases substantially from method one to four.

Poulos (2000) compared the capabilities of the various methods, concluding that no single test is able to satisfactorily supply all the information needed to the designer.

It is evident that the integrity test is essentially a structural investigation tool for bored cast *in situ* piles; it may be relevant to geotechnical design only where the detection of structural defects can be a valid support to the interpretation of the results of other test methods. Accordingly, in the following paragraphs attention will be focused on the more significant rapid and dynamic tests, with comparisons to the static load test.

7.3.2 Experimental layout

In a high strain dynamic load test a pile-driving hammer or a large drop-weight strikes blows down the pile, while the axial strain and the acceleration of the head of the pile are recorded. The raw data are logged during the tests at a very high frequency and converted to force F and velocity v or displacements w. The acceleration is usually integrated to obtain velocity or displacement while the axial strain is converted into the force F by multiplying the measured value by the cross-sectional stiffness EA of the pile.

In the portion of the pile above the ground surface the stress wave travels down unimpeded; the force and the velocity are directly proportional:

$$F = Zv = \frac{EA}{c}v$$

where Z is the pile impedance and c is the wave velocity in the pile.

Table 7.1 Dynamic vs. static load tests on a pile

Methods	Typical peak strain in pile	Typical force duration	Results
1 Static	$1000\,\mu\varepsilon$	10^6–$10^7\,\mathrm{ms}$	Bearing capacity and load–settlement curve
2 Statnamic or "rapid"	$1000\,\mu\varepsilon$	50–$200\,\mathrm{ms}$	Bearing capacity and load–settlement curve
3 Dynamic	500–$1000\,\mu\varepsilon$	5–$20\,\mathrm{ms}$	Bearing capacity and possibly load–settlement curve
4 Integrity	5–$10\,\mu\varepsilon$	0.5–$2\,\mathrm{ms}$	Structural defect in the pile

In the portion of the pile below the ground surface any resistance to movement of the pile will generate an upward travelling reflected wave propagating back towards the pile head. In other words, changes in the pile impedance are expected due to the shaft friction, and to any significant change in pile cross section. Simultaneous records of strain and acceleration (converted to velocity and force) allow the separation of downward and upward components of the stress wave. A scheme of the dynamic load test on a pile is reported in Figure 7.18.

The typical layout of a Statnamic or rapid load test is reported in Figure 7.19. A more or less fast burning fuel is used to accelerate a mass away from the pile, thereby loading it in reaction. Typically the mass is accelerated up to 20 G, to reduce the mass to only 5% of that needed to apply an equivalent static load. In the early version of the test the mass was left free to fall on a sand bed on the top of the pile. More recently a hydraulic piston is used to support the free falling mass (Figure 7.20). This improvement allows the use of the Statnamic method to test piles under cyclic vertical load and, less frequently, under horizontal load.

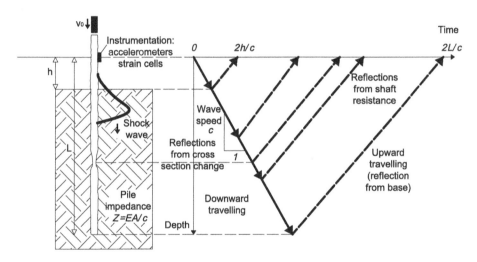

Figure 7.18 Schematic of stress wave during a dynamic load test (after Randolph 2003).

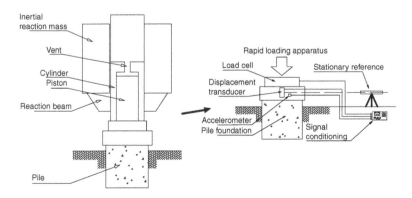

Figure 7.19 Schematic layout of a Statnamic load test on a pile.

Figure 7.20 Picture of a Statnamic vertical load test (3 MN maximum load).

The displacement and the load or the acceleration at the pile head are recorded also in this test. As reported in Figure 7.19, a laser displacement sensor and a strain gauge load cell are used to measure the applied load and the downward displacement. Sometimes accelerometers or strain gauges are also placed either on top of the pile or along the pile shaft; in the latter case the test provides valuable information on the load transfer along the shaft.

Load duration for both dynamic and rapid load tests is such that the soil resistance depending on the inertial forces of the soil and the rate dependency of the soil strength cannot be neglected. The opposite is true only for static load tests.

In terms of drainage conditions, a comparison between the loading duration and the time needed for the soil to fully consolidate around a loaded pile shows that both dynamic and Statnamic load tests must be considered as undrained tests in all types of soil, while, on the contrary, the static load test can be considered as a drained one.

Finally, it may be observed that both the high energy dynamic test and the rapid (or Statnamic) test have the definite advantage of reproducing very nearly the scheme of the ideal load test (ILT), as defined in §7.1.4.

7.3.3 Dynamic tests and stress wave theory

For the purpose of using them in design, the results of dynamic load tests on piles have to be transformed into an equivalent static load–settlement curve up to failure. This is particularly cumbersome and uncertain for the dynamic load test proper, while for the rapid or Statnamic test significant simplifications are provided by the longer duration (lower frequency) of the load application.

As illustrated in Figure 7.21, the range of natural frequencies of a typical pile–soil system is intermediate between the loading frequency of high strain dynamic tests and that of Statnamic tests. For this reason the Statnamic test, as the traditional static load test, does not generally need to be interpreted through the application of the stress wave theory to the pile while this is mandatory for high energy dynamic testing.

Due to the variety of soil conditions, geometry of piles and Statnamic procedures, this statement can better be specified using some quantitative parameter. To this aim Middendorp and Bielefeld (1995) defined a wave number, $N_w = cT/L$, where T is the load duration, c the wave velocity in the pile and L its length. There is a broad theoretical and experimental evidence that stress wave phenomena in a pile body can be neglected for $N_w > 12$.

The wave velocity c in a concrete pile is around 4000 m/s and in a steel pile 5200 m/s. For a Statnamic load test typical value of T is not less than 0.15. If a pile is 10 m long the wave number N_w ranges between 40 and 52 depending on the pile material: the stress wave phenomena down the pile body can be thus neglected. In the case of a very long pile, for instance 50 m or longer, the wave number N_w ranges between 8 and 10.4 and the interpretation of the Statnamic test should include the stress wave analysis.

The interpretation of the test without an analysis of the stress wave propagation in the pile body is often associated to a pile modelled as rigid body. This can be misleading, since significant strains develop in the pile during a Statnamic load test. In fact, for values of $N_w > 12$, the signals along the pile shaft are substantially all in phase, as it occurs in the extreme case of the static load tests. If on the contrary $N_w < 12$, the above statement does not hold true; this occurs for very long piles during rapid load tests, and for dynamic high energy tests whatever the pile.

To support these statements, Figure 7.22 reports the results of a Statnamic and a dynamic load test on a bored cast *in situ* pile with a diameter $d = 1.07$ m and a length $L = 27.3$ m (Pando *et al.* 2000). The displacement at the top and bottom of the pile,

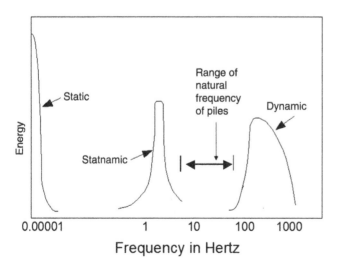

Figure 7.21 Typical frequency of load testing methods compared to the natural frequency of pile–soil system (after Bermingham and Janes 1989).

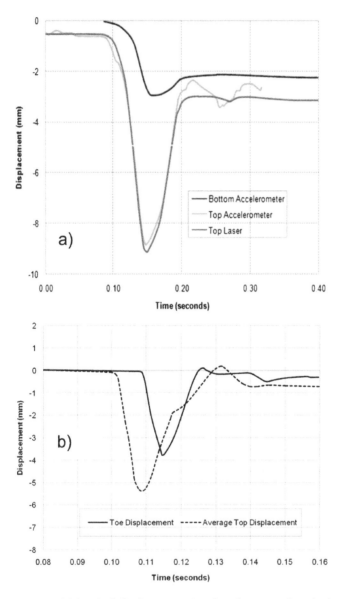

Figure 7.22 Typical displacement signal at the top and at the bottom of a pile during (a) a Statnamic load test and (b) a dynamic load test.

obtained by double integration of the measured acceleration, is reported; it can be clearly seen that in the Statnamic test (Figure 7.22.a) the displacement at the top and bottom of the pile are substantially in phase, while the same does not occur during the dynamic load test (Figure 7.22.b). The displacement at the top of the pile obtained by direct laser measurement in the Statnamic test are also reported; there is a satisfactory agreement between the direct measurement and the double integration of the acceleration.

For large values of N_w the time needed for the compression wave to travel downward and back is much smaller than the duration of loading; during the Statnamic tests, hence, the pile usually does not experience significant tensile stresses. On the contrary during a dynamic load test, significant tensile stresses are developed along the pile shaft and structural damages are possible. This is one of the reasons why prefabricated high quality reinforced concrete piles or steel piles are more often subjected to dynamic testing than ordinary cast *in situ* bored piles; in fact dynamic tensile forces which develop during driving operations are always accounted for at the design stage of this kind of pile. Furthermore, dynamic tests are usually executed for such piles for the availability of the current driving equipment. The exact knowledge of the cross section of prefabricated piles makes easier the interpretation of dynamic test results while, for bored cast *in situ* piles, structural defects could cause misinterpretation of the test. The structural defects and the ordinary load transfer along the pile shaft are in fact practically undistinguishable causes of changes in pile impedance along its axis.

7.3.4 Results and interpretation

The interpretation of a dynamic load test, aimed at deriving an equivalent static load–displacement relationship, is a rather complex procedure and its full description is outside the scope of this book. In §4.6 a brief outline of the phenomena occurring in a pile during driving has been provided; the stress wave equation is the basis for the interpretation of high strain dynamic testing (Figure 7.18). The most widespread approach is by computer simulation of the pile–soil interaction, varying the soil parameters until an acceptable match between the computed and recorded stress wave signals is achieved. Since the solution is not unique a number of different distributions of soil parameters gives a satisfactory fit to the measured response; a good deal of subjective judgement is anyway required. Furthermore, according to Randolph (2003) most of the commercial software currently available still use somewhat simplistic soil models.

For the Statnamic test, on the contrary, simplified methods are often adequate to derive the static equivalent response of the pile. The Unloading Point Method (UPM) (Middendorp 1993), is based on the equation:

$$F_{statnamic} = F_{static} + m_{pile} \cdot \frac{d^2 w_{pile-head}}{dt^2} + C \cdot \frac{dw_{pile-head}}{dt} \tag{7.3}$$

The equation implies that the pile body can be treated as a unique mass subjected to a given acceleration, usually measured or derived at the pile head. Viscous effects are lumped into the parameter C. Both inertial and viscous effects are summed to the static equivalent force to give the force $F_{statnamic}$ recorded during the Statnamic load test.

The UPM first identifies the point where the pile has zero velocity (unloading point) and assumes that the pile resistance at this point is equivalent to the static pile resistance if the inertia forces are negligible. By considering the pile resistance between the peak applied load and the unloading point a damping constant (Middendorp 1993) is found which is used to remove the rate dependent component of the rapid load test.

Many experimental results show that the accuracy of this simple method can be significantly improved by direct measurements of accelerations at both the pile head and at the pile toe. In fact, even if the two signals are generally in phase, they have different amplitude since the peak acceleration is a decreasing function of the depth along the pile shaft; the inertial force is hence overestimated using only the acceleration at the pile head.

Brown and Hyde (2008) propose a more complex formulation of Eq. 7.3 for Statnamic tests of piles in clayey soil.

Many experimental investigations show that the load–settlement response directly observed in rapid load tests for piles in coarse-grained soil or for rock-socketed piles is very similar to the load–settlement response by static load tests. In Figure 7.23 (Justason *et al.* 2000), for instance, the results of a Statnamic load test are compared to those of a static maintained load test (with three cycles). The pile was a 600 mm square prestressed concrete pile 10.5 m long, driven in sandy soils.

In Figure 7.23a the load–displacement curve recorded in the Statnamic test is converted into the static equivalent curve using the UPM.

In Figure 7.23b this curve is compared to that of the three static load cycles. A rather good agreement is observed, both in terms of stiffness and (extrapolated) bearing capacity. In Figure 7.23c, the main ingredient to derive by the UPM, the equivalent static curve, is plotted against the time.

7.3.5 Concluding remarks

Dynamic and rapid (or Statnamic) load tests are to be considered as Ideal Load Tests, according to the definition given in §7.1.4, and provide a quick and cost-effective

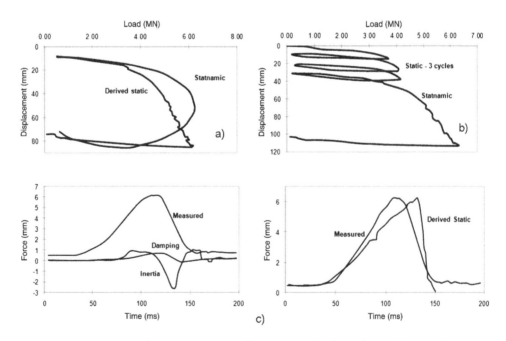

Figure 7.23 Comparison between a static and a Statnamic load test for a pile in sandy soil.

method of verifying that the working piles meet a performance specification. The methods do have however certain limitations, such as:

1   The dynamic bearing capacity may be significantly greater than the static, due to inertial and rate effects (damping). This is particularly true for dynamic testing where the accelerations induced in the pile are much higher than in a rapid load test; for this reason, particularly in fine-grained soils, the dynamic test seems not suited to determine the static bearing capacity.
2   The interpretation of the dynamic test for bored cast *in situ* piles can be very cumbersome because the stress wave matching analysis cannot properly distinguish between changes in the dynamic impedance due to structural defects of the pile shaft and load transfer to the soil along the shaft. Integrity testing can be coupled to dynamic load tests in order to reduce the uncertainties about structural issues.
3   The rapid or Statnamic load tests can be usually interpreted according to simple method as the UPM and neglecting the stress wave propagation. For very long piles and particularly in clay, however, the simple methods can be inaccurate. In such cases the same interpretation techniques used for high energy dynamic tests should be adopted.
4   Apart from rapid load tests on piles in very coarse-grained soil the predicted load–displacement curve will not take into account any consolidation or creep effects.

Notwithstanding these limitations both methods are actually widespread all over the world. The rapid load test, in particular, has gained in the last decade a significant part of the total market of pile load tests in the US and in Japan. Codes of practice and guidelines for rapid load test have been issued by institutions as ASTM, the US Transportation Research Board and the Japanese Geotechnical Society.

7.4 Horizontal load test

7.4.1 Introduction

Horizontal load tests on foundation piles are by far less common than vertical ones. As for vertical tests (§7.1.1), they can be subdivided into *design* and *proof load test;* unlike vertical tests, in practice the former is more usual. Being a destructive test usually implying the collapse of the structural section, a horizontal load test has to be carried out on a purposely installed test pile, which does not belong to the foundation. The aim of a design load test is to determine, at the design stage, the bearing capacity of the pile and its load-horizontal displacement relationship; if the pile shaft is properly instrumented, it allows the determination of the transfer curves of the side pressure (the so-called *p–y* curves).

7.4.2 Equipment and procedure

The test equipment and procedure are similar to those described for vertical load tests. The load is applied by means of a hydraulic jack; a support to hold the jack horizontally against the test pile has to be arranged. The reaction to the jack is usually provided by another pile or by a pair of piles connected by a horizontal beam. The applied load is measured by means of a load cell.

The horizontal displacement of the head of the test pile is measured by dial gauges or displacement transducers (or both), fixed to the pile head and contrasting on one or two reference beams, anchored to supports sufficiently spaced from the test pile.

During a horizontal load test the head of the pile is usually free to rotate, while the piles belonging to the foundation are connected to a cap representing a significant constraint. The rotation of the head of the pile during the test can be measured independently using tilt sensors, accelerometers probes or simply measuring horizontal displacement at two different levels above the loading jack.

The amount and value of information that can be obtained by a horizontal pile load test increase substantially if, in addition to the displacement and rotation of the pile head, the horizontal displacement along the pile shaft is measured; the usual way is to install an inclinometer tube along the axis of the pile. Alternatively the strains along two opposite peripheral fibres can be measured at various depths along the pile shaft, fitting strain gauges to the reinforcement bars. In both cases the curvature of the pile axis is obtained, allowing the determination of bending moments, shears and soil pressures by repeated integrations.

The most common test procedure is the so-called maintained load test, in which the load is applied in steps and each load step is kept constant for a certain time while the displacement is measured. It is suggested to adopt the same type of procedure presented for vertical load tests.

7.4.3 Interpretation

As recalled above, the principal aim of a horizontal pile load test is the determination of the horizontal bearing capacity, but the load–displacement relationship is also of some interest.

In both cases the degree of constraint at the pile head should be taken into proper account. As stated before, during the test usually the head of the pile is free to rotate while under service load the pile cap inhibits, at least partially, such a rotation and induces a bending moment at the pile head.

It could be easily shown that both the bearing capacity and the horizontal stiffness of a horizontally loaded pile with constrained head are significantly larger than the corresponding values for a pile with the head free to rotate.

For a deeper insight into the matter, reference can be made to Chapter 8 for the bearing capacity problem. In terms of load–displacement response, in Figure 7.24 the load displacement relationship and the profile of the displacement at a given load level, computed via a non-linear BEM analysis, are compared for the same pile in the same subsoil condition but with reference to the two different schemes of (i) pile head free to rotate and (ii) fixed-head pile. The response of the fixed-head pile is definitely stiffer than that of the free-headed pile; around a likely service load the stiffness of the former is at least twice that of the latter.

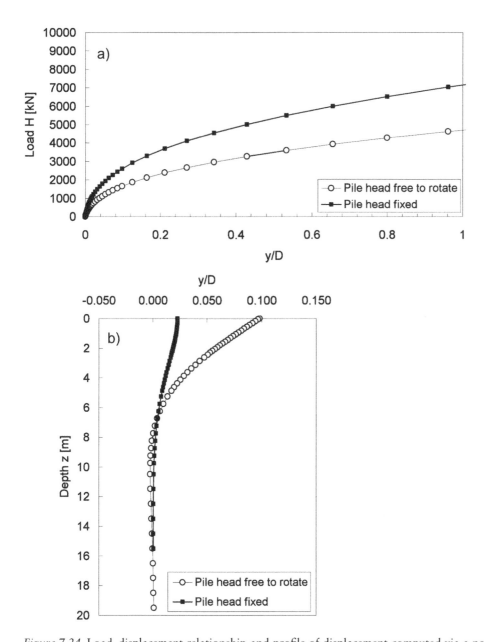

Figure 7.24 Load–displacement relationship and profile of displacement computed via a non-linear BEM analysis.

Part III

Present practice of piled foundations design under horizontal loads

8 Bearing capacity under horizontal load

8.1 Introduction

There are many examples of foundations subjected to significant horizontal forces and moments, such as marine and offshore structures, retaining structures and tall buildings under wind or seismic loading, and there are significant differences between the behaviour of piles under horizontal and vertical load. Under vertical axial load the structural section of the pile is subjected to confined compression; the stress is generally much lower than the strength of the material (wood, steel, concrete) of the pile; the failure, if any, occurs at the interface between the pile and the soil and the structural section of the pile does not raise significant design problems. Under lateral load, on the contrary, the pile is subjected to bending and shear, and the behaviour of its section is a major component of the pile response. The behaviour of a vertically loaded pile, and in particular its bearing capacity, depends essentially on the characteristics of the soil immediately adjacent to the shaft and below the base; in these zones the pile installation produces significant variations of the state of stress and soil properties. As has been shown in Chapter 4, the behaviour of a vertically loaded pile, and in particular its bearing capacity, is hence markedly affected by the installation procedures. Under horizontal load the pile–soil interaction is confined to a volume of soil close to the surface, a major part of which is not influenced by the pile installation. Accordingly, the behaviour of a laterally loaded pile is usually considered not to be significantly affected by the installation technique.

The available full-scale experimental evidence on piles under horizontal load is less exhaustive than for vertical loading. In particular, the values of bending moment in the pile shaft may be obtained only in an indirect way, and are affected by significant uncertainties. The characterization of the very shallow soils controlling the behaviour of the pile is seldom satisfactory. Finally, full-scale observations of the behaviour of pile groups under horizontal load are scarce.

8.2 Bearing capacity of the single pile

8.2.1 General

As with vertical loading, consideration must be given in design to both the ultimate lateral resistance of piles and the lateral deflection under the design service loadings. It is not common that the ultimate lateral resistance, or horizontal bearing capacity, is the governing factor in design; nevertheless, it is important to consider the ultimate lateral capacity of the pile and the ultimate lateral resistance of the soil, since they are also important components of a non-linear analysis of lateral response.

The calculation of the ultimate capacity of a pile under lateral loading requires a specification of the distribution of the ultimate lateral pile–soil pressure with depth, the structural strength of the pile in bending and the postulated failure mode of the pile–soil system.

The problem has been treated by a number of authors (Reese *et al.* 1974; Meyerhof 1995). The classical work in this area, however, has been published by Broms (1964a, 1964b), and his approach continues to be widely adopted in practice.

Let us consider a pile in a homogeneous soil; the horizontal stress at the pile–soil interface has initially an axially symmetric distribution and hence a resultant equal to zero (Figure 8.1). If we imagine that the pile undergoes a horizontal displacement δ within the soil, the interface stress changes as shown in Figure 8.1, with generation of shear stress in addition to normal stress, and its resultant p is a force per unit length of the pile, having the direction of the relative displacement pile–soil and opposite sign.

Broms assumes the p–δ relationship to be rigid-perfectly plastic; in other words, the ultimate resistance of the soil is entirely mobilized for any value of the displacement δ and keeps constant with increasing displacement (Figure 8.2a). Broms assumes also that p is independent of the shape of the pile section, but depends only on its width (diameter of a circular section, width in the direction normal to the movement for any other shape). The structural behaviour of the pile section is assumed to be rigid-perfectly plastic also, in the sense that the elastic rotation may be neglected until the bending moment attains the yield value M_y at a certain depth. At this stage a plastic hinge develops, and the rotation goes on indefinitely under a constant moment M_y (Figure 8.2b).

Considering again a horizontal displacement of the pile, the profile of the ultimate soil resistance adopted by Broms is depicted in Figure 8.3. For an undrained analysis

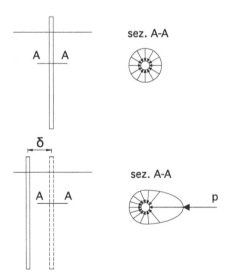

sez. A-A

sez. A-A

Figure 8.1 Pile-soil interaction.

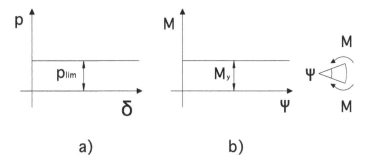

Figure 8.2 Rigid-perfectly plastic behaviour of (a) the soil and (b) the pile.

in fine-grained soils and in terms of total stress (homogeneous "cohesive" soil characterized by an undrained strength c_u), $p = 2c_u d$ at the surface, where the soil may
flow up toward the free surface. From a depth $z = 3d$ downward the soil is forced to
flow horizontally, and the ultimate resistance is $p = 8$–$12c_u d$ (Figure 8.3a). To make
the analysis easier, Broms suggests to adopt the simplified distribution shown in
Figure 8.3b.

For a drained analysis in coarse-grained soils and in terms of effective stress
(homogeneous "cohesionless" soil characterized by a friction angle φ' and a unit
weight γ or γ') Broms assumes that the ultimate lateral soil resistance increases linearly with depth according to the equation:

$$p = 3k_p \gamma dz \tag{8.1}$$

Barton (1984), Fleming *et al.* (1985), Kulhawy and Chen (1993), Russo and Viggiani (2008) claim that the Broms' ultimate resistance for cohesive soils is but slightly

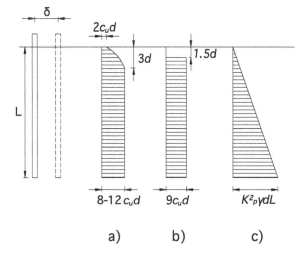

Figure 8.3 Profile of the ultimate soil resistance assumed by Broms (1964a, 1964b):
(a) cohesive soil experimental; (b) cohesive soil simplified; (c) cohesionless soil.

conservative, while for cohesionless soils it generally underestimates the observed value by a significant amount. Barton (1984) suggested using, instead of Eq. 8.1, the following relationship:

$$p = k_p^2 \gamma dz \tag{8.2}$$

For relatively loose soils ($\varphi \leq 32°$) the two relations give very similar results; for denser soils, Barton's suggestion gives values of the lateral soil resistance significantly higher than Broms' one. In the following development, the Barton equation (Figure 8.3c) will be used.

8.2.2 Free-head pile, cohesive soil

Let us consider a pile in a homogeneous cohesive soil, with the head free to rotate and subjected to a horizontal force H and a moment He. Failure may occur following one of the two mechanisms reported in Figure 8.4. In Figure 8.4a (so-called "short pile" mechanism) the maximum bending moment in the pile M_{max} is smaller than the yield moment M_y of the pile section, and hence a rigid rotation of the pile

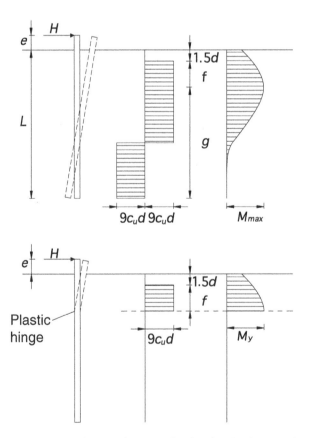

Figure 8.4 Failure mechanisms for free-head piles in cohesive soils: (a) "short" pile; (b) "long" pile.

occurs; the ultimate value of the applied horizontal force $H_{\lim}$ (horizontal bearing capacity) depends only on the geometry of the pile (values of L, d, e) and on the undrained strength of the soil.

When the maximum bending moment in the pile exceeds the yield value M_y, on the contrary, the failure mechanism is that shown in Figure 8.4b ("long pile"). A plastic hinge is formed, and the horizontal bearing capacity will depend on M_y, besides the geometry and the soil strength.

In the case of short pile, at a depth $z=(1.5d+f)$ the moment is maximum and hence the shearing force is equal to zero. Horizontal equilibrium of the pile above this section requires that:

$$f = \frac{H_{\lim}}{9c_u d} \tag{8.3}$$

The equilibrium of the whole pile around the point of maximum moment gives:

$$9c_u d \frac{g^2}{4} = H_{\lim}\left(e+1.5d+f\right)-9c_u d \frac{f^2}{2} \tag{8.4}$$

Taking into account that $L=1.5d+f+g$, Eqs 8.3 and 8.4 allow the determination of $H_{\lim}$ which can be expressed in the following dimensionless form:

$$\frac{H_{\lim}}{c_u d^2} = -9\left(1.5+\frac{L}{d}+\frac{2e}{d}\right)+9\sqrt{2\left(\frac{L}{d}\right)^2+4\left(\frac{e}{d}\right)^2+4\frac{Le}{d^2}+6\frac{e}{d}+4.5} \tag{8.5}$$

Eq. 8.5 is reported in Figure 8.5.

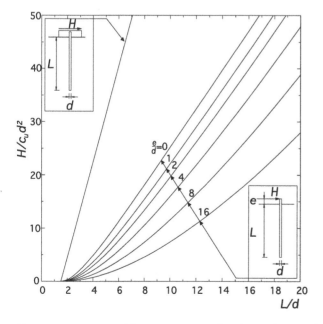

Figure 8.5 Horizontal bearing capacity for short piles in cohesive soils.

For the mechanism of short pile to be valid, it is to be checked that $M_{max} \leq M_y$. The dimensionless expression of M_{max} is:

$$\frac{M_{max}}{c_u d^3} = \frac{H_{lim}}{c_u d^2}\left(\frac{H_{lim}}{18 c_u d^2} + \frac{e}{d} + 1.5\right) \tag{8.6}$$

Inserting into Eq. 8.6 the expression of $\dfrac{H_{lim}}{c_u d^2}$ given by Eq. 8.5, the values of $\dfrac{M_{max}}{c_u d^3}$ reported in Figure 8.6 as a function of L/d and e/d are obtained.

If $M_{max} \leq M_y$, the pile under examination is actually short and the horizontal bearing capacity is given by Eq. 8.5. If, on the contrary, $M_{max} > M_y$, the pile is actually a long one. Eq. 8.3 is still valid; the rotational equilibrium of the pile above the plastic hinge requires that:

$$M_y = H_{lim}\left(e + 1.5d + 0.5f\right) \tag{8.7}$$

Combining Eqs 8.3 and 8.7 one obtains:

$$\frac{H_{lim}}{c_u d^2} = -9\left(1.5 + \frac{e}{d}\right) + 9\sqrt{\left(\frac{e}{d}\right)^2 + 3\frac{e}{d} + \frac{2M_y}{9 c_u d^3} + 2.25} \tag{8.8}$$

Eq. 8.8 is reported in Figure 8.7. The horizontal bearing capacity of a long pile depends on c_u, d, e and M_y but is independent of L. This apparently paradoxical result may be explained by the fact that the pile must be long enough to establish the mechanism of a long pile. Any further increase of the pile length above that value

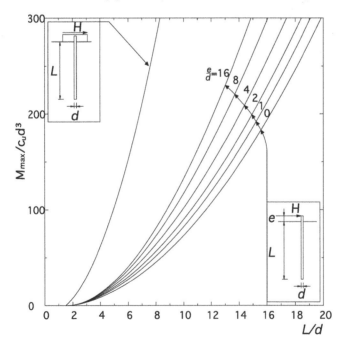

Figure 8.6 Maximum bending moment for short piles in cohesive soils.

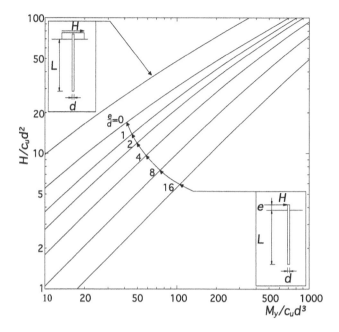

Figure 8.7 Horizontal bearing capacity for long piles in cohesive soils.

has no effect on the horizontal bearing capacity. The minimum length required for a long pile mechanism may be obtained entering the graph in Figure 8.6 with the value of $\dfrac{M_y}{c_u d^3}$.

Given a certain pile (l, d, e, M_y) and a certain soil (c_u), the actual value of the horizontal bearing capacity will be the smaller of the two values given by Eqs 8.5 and 8.8.

8.2.3 Fixed-head pile, cohesive soils

In practice, the case of a free head is not frequent, the pile head being generally connected to a cap and a superstructure which hinder totally or partially the rotation while permitting horizontal displacement. We will examine the case of a pile connected at the ground surface ($e=0$) to a structure capable of preventing any rotation. The results that will be presented are thus dependent on this hypothesis which, anyhow, is often a satisfactory model of the actual situation.

The possible failure mechanisms are in this case three (Figure 8.8): "short", "intermediate" and "long" pile.

For the short pile (Figure 8.8a) the horizontal equilibrium requires that:

$$H_{\lim} = 9c_u d\left(L-1.5d\right)$$

and hence:

$$\frac{H_{\lim}}{c_u d^2} = 9\left(\frac{L}{d}-1.5\right)$$

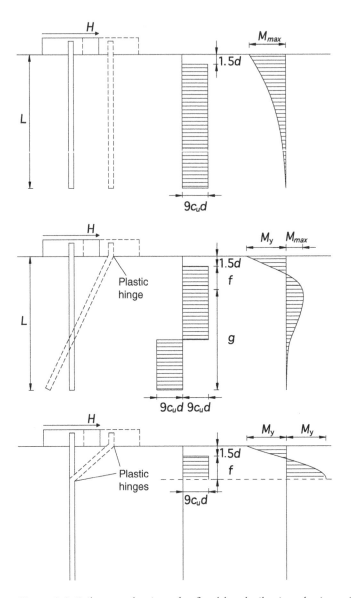

Figure 8.8 Failure mechanisms for fixed-head piles in cohesive soils: (a) "short" pile; (b) "intermediate" pile; (c) "long" pile.

This expression is reported in Figure 8.5, together with the curves referring to the free-headed piles. $H_{\lim}$ depends again only on L, d and c_u. It has to be checked that $M_{max} \leq M_y$. To this aim we evaluate:

$$M_{max} = H_{\lim}\left(0.5L + 0.75d\right)$$

Inserting the above value of $H_{\lim}$ given by Eq. 8.9, we obtain:

$$\frac{M_{max}}{c_u d^3} = 4.5\left(\frac{L}{d}\right)^2 - 10.125 \tag{8.9}$$

This expression is plotted in Figure 8.6 together with the corresponding curves referring to the free-headed piles.

For the intermediate pile (Figure 8.8b) a plastic hinge develops at the connection between the pile and the cap. The horizontal equilibrium of the pile above the section of maximum moment is again expressed by Eq. 8.3. The rotational equilibrium around the plastic hinge gives:

$$M_y + 9c_u d \frac{g^2}{4} - 9c_u df\left(\frac{f}{2} + 1.5d\right) = 0 \tag{8.10}$$

Combining Eqs 8.3 and 8.10 one obtains:

$$\frac{H_{lim}}{c_u d^2} = -9\left(\frac{L}{d} + 1.5\right) + 9\sqrt{2\left(\frac{L}{d}\right)^2 + \frac{4M_y}{9c_u d^3} + 4.5} \tag{8.11}$$

This expression is reported in Figure 8.9. The horizontal bearing capacity is a function of L, d, c_u and M_y. The different curves, each referring to a value of $\dfrac{M_y}{c_u d^3}$, are bounded on the left by the equation, valid for short piles. Toward the right, the curves are valid until the length of occurrence of the second plastic hinge is reached.

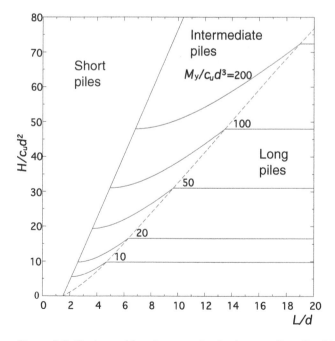

Figure 8.9 Horizontal bearing capacity for intermediate fixed-head piles in cohesive soils.

For the long pile (Figure 8.8c), the horizontal equilibrium of the pile between the two plastic hinges gives:

$$H_{\text{lim}}\left(1.5d + 0.5f\right) = 2M_y$$

Taking into account Eq. 8.3, one obtains:

$$H_{\text{lim}}^2 + 27c_u d^2 H_{\text{lim}} - 36c_u dM_y = 0$$

and finally:

$$\frac{H_{\text{lim}}}{c_u d^2} = -13.5 + \sqrt{182.25 + 36\frac{M_y}{c_u d^3}} \qquad (8.12)$$

This expression gives the upper bound of the curves in Figure 8.9; for higher L/d values, they become horizontal since the horizontal bearing capacity of long piles is independent of L. Eq. 8.12 is plotted in Figure 8.7, together with the similar curves referring to the free-headed piles.

For each value of $\dfrac{M_y}{c_u d^3}$ Figure 8.9 allows the recognition of the ranges of L for short, intermediate and long piles respectively.

8.2.4 Free-head pile, cohesionless soils

Figure 8.10 reports the possible failure mechanisms for free-head piles in cohesionless soils; in this case also a short pile and a long pile mechanism may occur.

For a short pile, the centre of rotation is very close to the pile tip. In order to simplify the analysis and without a significant error, Broms suggests assuming that the rotation occurs around the tip of the pile; the ultimate soil reaction below the centre of rotation is assumed to be a concentrated force F (Figure 8.10a).

The rotational equilibrium around the pile tip requires that:

$$H_{\text{lim}}\left(e + L\right) = \frac{L^3}{6}k_p^2\gamma d$$

and hence:

$$\frac{H_{\text{lim}}}{k_p^2\gamma d^3} = \frac{1}{6}\frac{d}{e+L}\left(\frac{L}{d}\right)^3 \qquad (8.13)$$

Eq. 8.13 is plotted in Figure 8.11; as for cohesive soils, H_{lim} is a function of the shear strength of the soil (γ, k_p) and of the geometry of the pile (L, d, e). It remains to check that $M_{\text{max}} \leq M_y$. To this aim, we can note that the shearing force T along the pile shaft may be expressed as:

$$T = H_{\text{lim}} - k_p^2\gamma d\frac{z^2}{2}$$

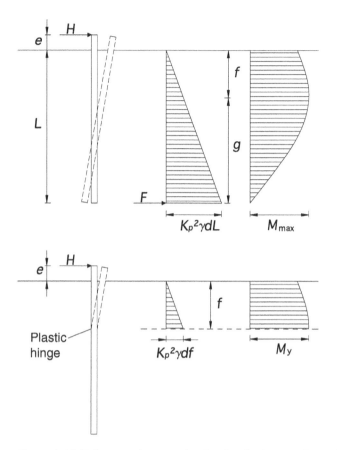

Figure 8.10 Failure mechanisms for free-head piles in cohesionless soils: (a) "short" pile; (b) "long" pile.

The depth of maximum moment f is that at which $T=0$; it is expressed:

$$f = \sqrt{\frac{2H_{\lim}}{k_p^2 \gamma d}}$$ (8.14)

The maximum moment may be expressed as:

$$M_{max} = H_{\lim}(e+f) - k_p^2 \gamma d \frac{f^3}{6} = H_{\lim}\left(e + \frac{2}{3}f\right)$$ (8.15)

Substituting the expressions 8.13 and 8.14 we obtain finally:

$$\frac{M_{max}}{k_p^2 \gamma d^4} = \frac{1}{6}\left(\frac{L}{d}\right)^2 \frac{d}{e+L}\left(\frac{e}{d} + \frac{2}{3}\frac{L}{d}\sqrt{\frac{1}{3}\frac{L}{e+d}}\right)$$ (8.16)

Eq. 8.16 is plotted in Figure 8.12. If $M_{max} \leq M_y$, the pile under examination is actually short and the horizontal bearing capacity is given by Eq. 8.13. If, on the

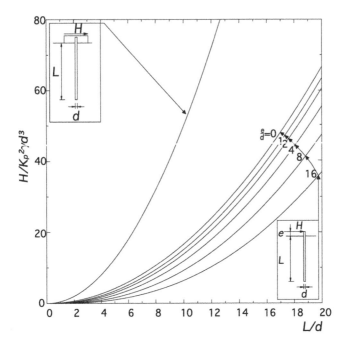

Figure 8.11 Horizontal bearing capacity for short piles in cohesionless soils.

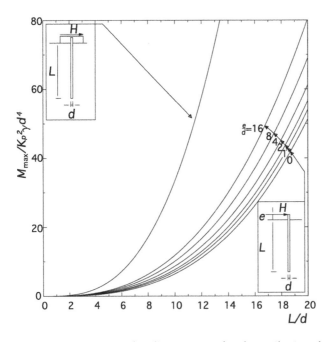

Figure 8.12 Maximum bending moment for short piles in cohesionless soils.

contrary, $M_{max} > M_y$, the pile is actually a long one. In this case, equating M_{max} to M_y one obtains from Eqs 8.14 and 8.15:

$$\frac{H_{\lim}}{k_p^2 \gamma d^3}\left(\frac{e}{d} + \frac{2}{3d}\sqrt{\frac{2H_{\lim}}{k_p^2 \gamma d}}\right) = \frac{M_y}{k_p^2 \gamma d^4} \tag{8.17}$$

This equation gives rise to the graph of Figure 8.13. As in cohesive soils, $H_{\lim}$ is independent of the length L of the pile, provided that the length is not smaller than that needed to the mechanism of long pile. The minimum length for a long pile mechanism may be obtained entering the graph in Figure 8.12 with the value of $\dfrac{M_y}{k_p^2 \gamma d^4}$.

Again, given a certain pile (l, d, e, M_y) and a certain soil (γ, k_p), the actual value of the horizontal bearing capacity will be the smaller of either values given by Eqs 8.13 and 8.17.

8.2.5 Fixed-head piles, cohesionless soils

The possible failure mechanisms and the corresponding distributions of soil reaction are reported in Figure 8.14.

For the short pile (Figure 8.14a), the horizontal equilibrium requires that:

$$H_{\lim} = \frac{k_p^2 \gamma d L^2}{2}$$

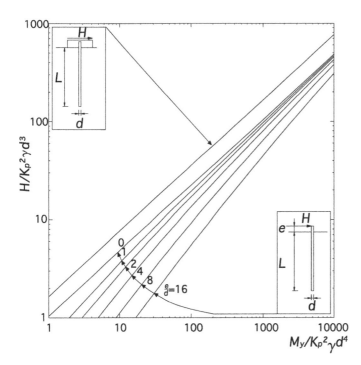

Figure 8.13 Horizontal bearing capacity for long piles in cohesionless soils.

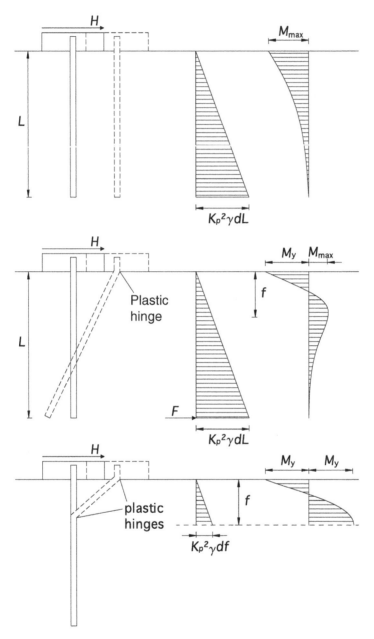

Figure 8.14 Failure mechanisms for fixed-head piles in cohesionless soils: (a) "short" pile; (b) "intermediate" pile; (c) "long" pile.

and hence:

$$\frac{H_{\lim}}{k_p^2 \gamma d^3} = \frac{1}{2}\left(\frac{L}{d}\right)^2 \tag{8.18}$$

Eq. 8.18 is plotted in Figure 8.11, together with the corresponding curves of the free-headed short pile. $H_{\lim}$ depends only on soil resistance (γ, k_p) and pile geometry (d, L); as usual, however, it is necessary to check that $M_{\max} \leq M_y$. To this aim it can be shown that:

$$M_{\max} = \frac{2}{3} H_{\lim} L$$

and hence:

$$\frac{M_{\max}}{k_p^2 \gamma d^4} = \frac{1}{3}\left(\frac{L}{d}\right)^3 \tag{8.19}$$

Eq. 8.19 is plotted in Figure 8.12.

For the intermediate pile (Figure 8.14b) a plastic hinge develops at the connection between the pile and the cap. The rotational equilibrium around the pile tip requires that:

$$H_{\lim} L - M_y - k_p^2 \gamma d \frac{L^3}{6} = 0$$

and hence:

$$\frac{H_{\lim}}{k_p^2 \gamma d^3} = \frac{M_y}{k_p^2 \gamma d^4} \frac{d}{L} + \frac{1}{6}\left(\frac{L}{d}\right)^2 \tag{8.20}$$

The horizontal bearing capacity is a function of L, d, γ, k_p and M_y. Eq. 8.20 is plotted in Figure 8.15, entirely corresponding to Figure 8.9; the graph in Figure 8.15 allows an immediate appreciation of the range of L/d corresponding to short, intermediate or long piles.

For the long pile, the rotational equilibrium of the pile between the two plastic hinges requires that:

$$\frac{2}{3} H_{\lim} f = 2M_y$$

Inserting the expression of f given by Eq. 8.14 one obtains:

$$\frac{H_{\lim}}{k_p^2 \gamma d^3} = \sqrt[3]{4.5\left(\frac{M_y}{k_p^2 \gamma d^4}\right)^2} \tag{8.21}$$

Eq. 8.21 is plotted in Figure 8.13 together with the similar curves of the free-headed long piles. The horizontal bearing capacity is a function of γ, k_p and d while it is

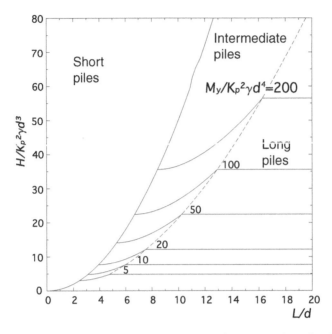

Figure 8.15 Horizontal bearing capacity for intermediate fixed-head piles in cohesionless soils.

independent of *L*. For long piles *L* does not appear explicitly, but it must have the minimum value reported in Figure 8.15 as the upper bound for intermediate piles.

8.2.6 Miscellaneous

The solutions provided in analytically closed form and represented in the graphs of the previous paragraphs are only valid for the simple cases of homogeneous cohesive soil with cohesion constant with depth, or homogeneous cohesionless soil either dry or with the groundwater table at the ground surface; the pile should be either free or fully restrained to rotate at the head. Broms' theory, however, may be applied to different subsoil profiles and boundary conditions, at the cost of finding a solution in each particular case. Some examples of such problems are reported in Figure 8.16.

8.3 Bearing capacity of the pile group

The horizontal bearing capacity of a pile group is generally smaller than the sum of the bearing capacities of the single piles; using the terminology already introduced for vertical loading, the efficiency of a group of piles in relation to horizontal loads is smaller than unity.

   The available experimental evidence is limited to small groups of piles or to laboratory tests on small scale models at natural gravity or in centrifuge. On this basis it appears that, when the spacing of piles is not less than six diameters in the direction of the horizontal load and of four diameters in the orthogonal direction, the efficiency tends to unity irrespective of the soil type. For the usual values of the spacing,

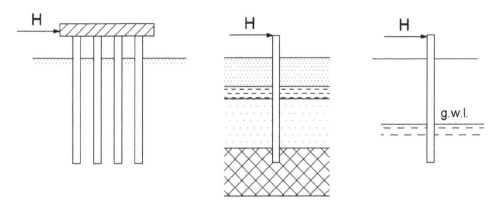

Figure 8.16 Examples of different problems that can be solved by means of Broms' theory.

(2.5 to 3 times the diameter) the efficiency may be conservatively assumed of the order of 0.5. For spacing between 2.5 and 6 (or 4), linear interpolation is suggested.

From the viewpoint of the bearing capacity, a layout with the piles placed normally to the direction of the horizontal load is to be preferred; in other words, given the number of piles in the group, a rectangular group is more efficient when the horizontal load acts parallel to the short side of the group.

The addition of inclined piles to a group is very effective in increasing both the stiffness and the bearing capacity of the group under horizontal load.

9 Displacements and bending moments

9.1 Single pile

9.1.1 Experimental evidence

As a first issue, the possible influence of the installation technique on the behaviour of laterally loaded piles will be investigated by reviewing the results of a small number of full-scale horizontal load tests on piles in the same site but with different installation techniques, reported in the literature.

As an example, Figure 9.1 reports the results of four load tests on prefabricated piles, carried out in the framework of the Arkansas River Project (Alizadeh and Davisson 1970). The subsoil at the site consists mainly of dense sand and the groundwater table is located very close to the ground level. Piles E7 and F7 were installed by jetting to 8 m and then driven to 15 m; piles E3 and F3 had been driven from the ground surface. The different installation techniques do not appear to affect the results; in this case the major factor seems to be the position of the test pile within the group.

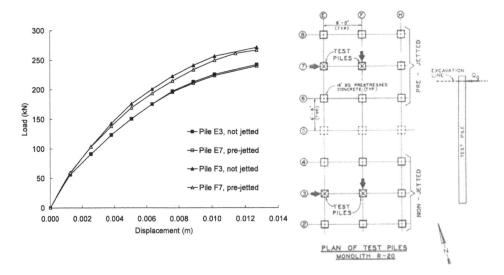

Figure 9.1 Horizontal load tests on four different piles in the same subsoil (data from Alizadeh and Davisson 1970).

Figure 9.2 reports the results of two load tests in sandy soils (Huang *et al.* 2001), in which the test piles differ both for the installation technique (a cast in place bored pile and a driven prefabricated one) and for the flexural stiffness of the section (the section of the bored pile is 8.7 times stiffer than that of the driven one). The ratio between the initial tangents to the load displacement curves is $K_c = 5.3$; with increasing load, the ratio increases further. Should the difference be due only to the different flexural stiffness, K_c should be in the range $\sqrt[5]{8.7} = 1.54 \leq K_c \leq \sqrt[4]{8.7} = 1.71$. (see Figure 9.12)

It appears, therefore, that in this case the installation technique has some influence on the behaviour of the piles; furthermore, unexpectedly, the bored pile is stiffer than the driven one, in spite of the fact that CPT carried out before and after pile driving show that the relative density of the soil has been increased by pile driving.

The load–displacement curves obtained by Mori (2003) testing two different piles in the same sandy gravel subsoil are compared in Figure 9.3. The "Tsubasa Pile" is a

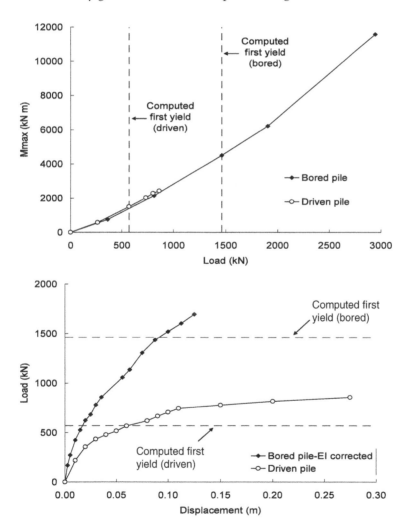

Figure 9.2 Horizontal load tests on a bored and a driven pile in the same subsoil (data from Huang *et al.* 2001).

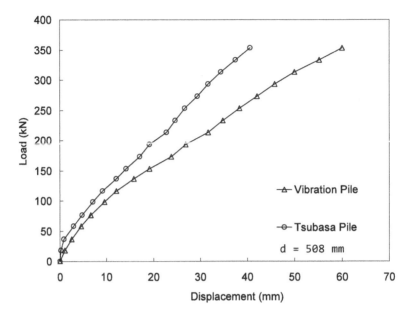

Figure 9.3 Horizontal load tests on a displacement screw pile (Tsubasa) and a bored pile
in the same subsoil (data from Mori 2003).

kind of displacement screw pile, the other an ordinary open-end steel pile driven by
vibration and removing the soil inside. The installation of the displacement screw
pile obviously densifies the surrounding soil; as expected, its stiffness is significantly
higher than that of the open-end driven pile.

Figure 9.4 reports the results of load tests on two piles (P4 and P5) bored through
a clayey fill (Callisto 1994). The pile P5 has a permanent steel casing in the upper
part of the shaft, conceived to decrease the negative friction; between the casing and
the surrounding soil there is a thin annulus (about 10 mm) filled with bentonite mud.
The initial tangent stiffness of the load–displacement curve of this pile, in spite of its
larger flexural stiffness due to the steel casing, is lower than that of pile P4; again,
the installation technique exerts an influence on the load–displacement response.

The horizontal bearing capacity of the two piles has been evaluated as 90% of the
asymptotic value of a hyperbola fitted to the data; it is equal to 815 MN for pile P4
and 1580 MN for pile P5. Such differences are accounted for by the different yield
moment of the two sections; therefore, in this case the installation technique does
not affect bearing capacity.

Summing up the above and other findings, it can be concluded that some influence of the installation technique has been observed, but the available data are scarce
and somewhat contradictory.

The deformations of the shaft of the piles P4 and P5 reported by Callisto (1994)
were monitored by means of strain gauges fitted to the reinforcement cage. From
these measurements, and taking into account the non-linearity of the moment-
curvature relation for the reinforced concrete section, the values of the bending
moment reported in Figure 9.5 have been obtained.

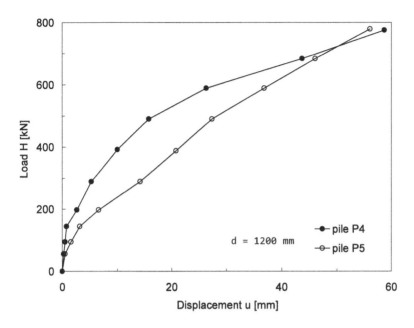

Figure 9.4 Horizontal load tests on two different piles in the same subsoil (data from Callisto 1994).

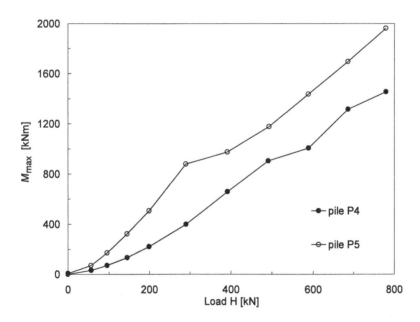

Figure 9.5 Load–maximum bending moment relationships (data from Callisto 1994).

Ruesta and Townsend (1997) report the results of a horizontal load test on a square reinforced concrete prefabricated pile, driven into a sandy subsoil (Figure 9.6). The test attained a displacement as high as 15% of the pile side (0.76 m); the test load has exceeded the value corresponding to the fissuring of concrete, and approached the horizontal bearing capacity, corresponding to complete yield of the section. The load–displacement curve is markedly non-linear from the very beginning; on the contrary, the load–maximum bending moment curve is nearly linear.

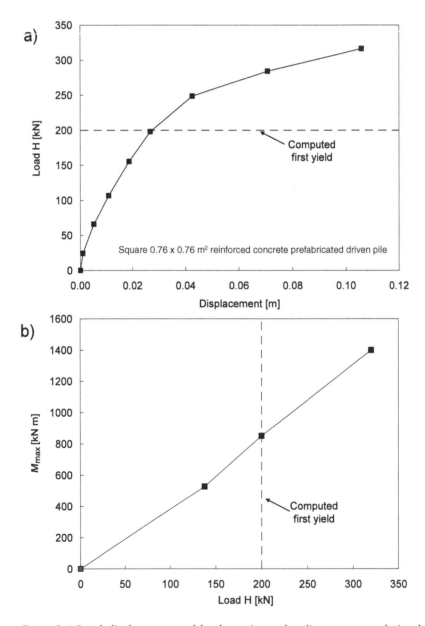

Figure 9.6 Load–displacement and load–maximum bending moment relationships (data from Ruesta and Townsend 1997).

Similar results have been obtained by Brown *et al.* (1987, 1988) for driven tubular piles both in coarse- and fine-grained soils (Figure 9.7). The same trend may be observed in Figures 9.2 and 9.5, and in quite a number of load tests.

It can be concluded that the lateral load-horizontal displacement relationship of a single pile is markedly non-linear, even at relatively low stress level; if a reliable prediction of displacement is critical, non-linearity has to be taken into account. On the contrary, it seems that the maximum bending moment is linked to the applied lateral

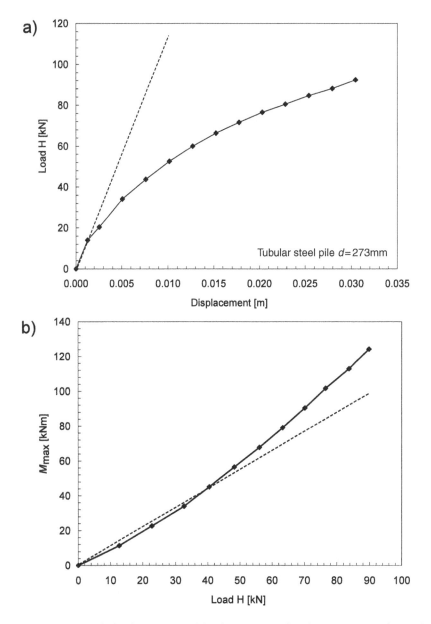

Figure 9.7 Load–displacement and load–maximum bending moment relationships (data from Brown *et al.* 1987).

load by a nearly linear relationship; this is an interesting observation if the main design issue is a structural one. For concrete piles non-linearity of the soil and the pile material add each other in determining the load–settlement response; on the contrary they compensate, at least partially, in determining the load-bending moment response. This is probably a factor of their observed response, while it does not apply for steel piles.

9.1.2 Winkler linear spring model

The methods for predicting the lateral deflection, rotation and bending moment of a laterally loaded pile model the soil either as a bed of horizontal independent Winkler springs (theory of subgrade reaction) or as a deformable continuum, solved by BEM or FEM.

Calling y the horizontal displacement of the pile at a depth z and P the reaction of the soil per unit length of the pile, and assuming $p = P/d$, the equation of the Winkler type model is:

$$p = k_h y \tag{9.1}$$

where $k_h [FL^{-3}]$ is the coefficient of horizontal subgrade reaction. Many authors prefer the parameter:

$$E_s = k_h d$$

which has the dimension of a modulus $[FL^{-2}]$ and is called the modulus of reaction of the soil. In fact, this parameter is less influenced than k_h by factors as the size of the pile, but in any case it is not a property of the soil, as it could be misinterpreted; accordingly, we will adopt here the definition in Eq. 9.1.

The foundation piles are usually "infinitely long" in the sense of Winkler model; the external loads are concentrated forces and moments at the top of the pile. In these conditions the simple Winkler model is known to give acceptable results. Furthermore, the state of stress and deformation in the pile is markedly affected by any variation of the soil characteristics along the pile axis; such variations are very frequent for piles, that cross different soil layers, and may be easily simulated by simply giving each layer a proper value of k_h and solving the equation by a finite difference or Finite Element Method.

In the case of relatively homogeneous subsoil, closed form solutions are available for the two simple cases of k_h constant with depth z and k_k linearly varying with depth with the equation:

$$k_h = n_h \frac{z}{d}$$

which is generally quoted from Matlock and Reese (1956). The case of constant k_h may model the case of an overconsolidated clay soil, while the model of linearly varying k_h is more suited for cohesionless soils or NC clays. The available solutions depend on the ratio between the length of the pile L and a characteristic length λ having the following expression:

$$\lambda = \sqrt[4]{\frac{4E_pJ}{k_h d}} \quad \text{for } k_h \text{ constant with depth}$$

$$\lambda = \sqrt[5]{\frac{E_pJ}{n_h}} \quad \text{for } k_h = n_h \frac{z}{d}$$

where E_p and J are respectively the Young modulus of the pile material and the moment of inertia of its section.

Some solutions for displacement and rotation at the top of a pile either free or restrained to rotate are reported in Table 9.1.

For a pile of finite length and stiffness $(2 < L/\lambda < 4)$ free to rotate and subjected to a horizontal force H and a moment M at the top, the following formulas may be written:

- Deflection $y = A_y \dfrac{H\lambda^3}{E_pJ} + B_y \dfrac{M\lambda^2}{E_pJ}$

- Rotation $\theta = A_\theta \dfrac{H\lambda^2}{E_pJ} + B_\theta \dfrac{M\lambda}{E_pJ}$ \hfill (9.2)

Table 9.1 Solutions for displacement and rotation at the top of a pile

Free-head pile, horizontal force H and moment M at the top

		$k_h = \text{const.}; \ \lambda = \sqrt[4]{\dfrac{4E_pJ}{k_h d}}$	$k_h = n_h \dfrac{z}{d}; \ \lambda = \sqrt[5]{\dfrac{E_pJ}{n_h}}$
Rigid pile $\dfrac{L}{\lambda} < 2$	y_o	$\dfrac{4H}{k_h dL} + \dfrac{6M}{k_h dL^2}$	$\dfrac{18H}{n_h L^2} + \dfrac{24M}{n_h L^3}$
	θ_0	$\dfrac{6H}{k_h dL^2} + \dfrac{12M}{k_h dL^3}$	$\dfrac{24H}{n_h L^3} + \dfrac{36M}{n_h L^4}$
Infinitely long pile $\dfrac{L}{\lambda} > 4$	y_o	$\dfrac{2H}{k_h d\lambda} + \dfrac{2M}{k_h d\lambda^2}$	$\dfrac{2.40H}{n_h^{3/5}\left(E_pJ\right)^{2/5}} + \dfrac{1.60M}{n_h^{2/5}\left(E_pJ\right)^{3/5}}$
	θ_0	$\dfrac{2H}{k_h d\lambda^2} + \dfrac{4M}{k_h d\lambda^3}$	$\dfrac{1.60H}{n_h^{2/5}\left(E_pJ\right)^{3/5}} + \dfrac{1.74M}{n_h^{1/5}\left(E_pJ\right)^{4/5}}$

Pile head restrained to rotate, horizontal force H

Rigid pile	y_o	$\dfrac{H}{k_h dL}$	$\dfrac{2H}{n_h L^2}$
	M_o	$\dfrac{HL}{2}$	$\dfrac{2HL}{3}$
Infinitely long pile	y_o	$\dfrac{H}{k_h dL}$	$\dfrac{H}{k_h dL}$
	M_o	$H\lambda$	$0.93H\lambda$

- Bending moment $M = A_M H\lambda + B_M M$

- Shear force $T = A_T H + B_T \dfrac{M}{\lambda}$

The functions $A(L/\lambda, z/\lambda)$, $B(L/\lambda, z/\lambda)$ are reported in Figure 9.8.

If the pile is restrained to rotate at the top and subjected to a horizontal force H, the bending moment M that develops at the pile head to restrain the rotation may be computed by imposing that:

$$\theta(z = 0) = A_\theta(z = 0)\frac{H\lambda^2}{E_p J} + B_\theta(z = 0)\frac{M\lambda}{E_p J} = 0$$

and hence:

$$M(z = 0) = \frac{A_\theta(z = 0)}{B_\theta(z = 0)} H\lambda = C_{Mo} H\lambda \qquad (9.3)$$

The values of the coefficient C_{Mo} are reported in Table 9.2.

Once the top moment is known, the displacement, rotation, bending moment and shear force along the pile shaft may be obtained by Eq. 9.2. For ease of computation, and for a pile restrained to rotate at the top and loaded by a horizontal force H, we can define:

- deflection $y = C_y \dfrac{H\lambda^3}{E_p J}$

- rotation $\theta = C_\theta \dfrac{H\lambda^2}{E_p J}$

- bending moment $M = C_M H\lambda$
- shear force $T = C_T H$

The functions $C(L/\lambda, z/\lambda)$ may be easily deduced by knowing the functions $A(L/\lambda, z/\lambda)$ and $B(L/\lambda, z/\lambda)$ and the coefficient C_{Mo}. Just as an example in Figure 9.9 the values of the function C_y are plotted.

Some suggestion for the evaluation of parameters are reported in the following.

The model of k_h constant with depth is suited for relatively stiff cohesive soils. Broms (1964a) suggested connecting the value of the coefficient of subgrade

Table 9.2 Values of C_{Mo} (Eq. 9.3)

L/λ	C_{Mo}
2	−1.06
3	−0.97
4	−0.93
≥5	−0.93

reaction to the undrained soil modulus E_{50}, determined in undrained triaxial compression tests at a deviatoric stress level equal to half the failure one, via the expression:

$$k_h = 1.67 \frac{E_{50}}{d}$$

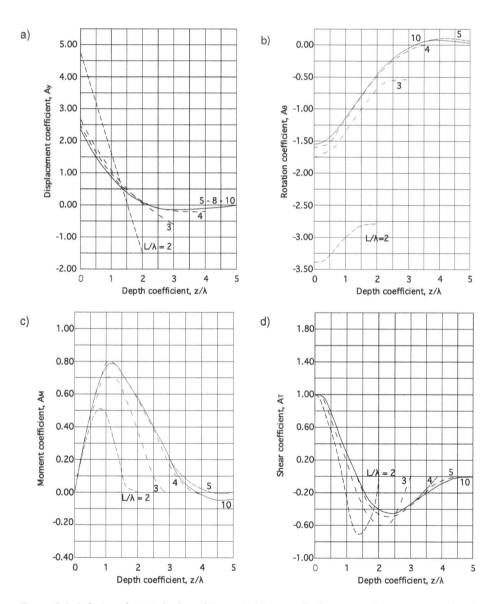

Figure 9.8 Solutions by Matlock and Reese (1956): (a) displacement, (b) rotation, (c) bending moment, (d) shear force for an horizontal head force; (e) displacement; (f) rotation; (g) bending moment; (h) shear force for a moment applied at the pile head.

continued

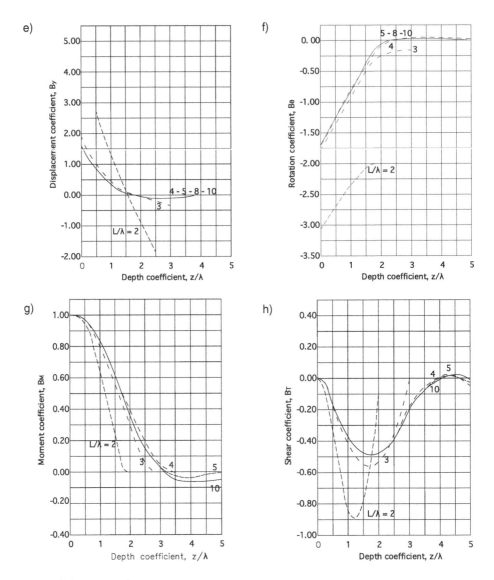

Figure 9.8 continued.

Being the undrained modulus for OC clays in the range 100 to 500 times c_u, it follows that:

$$k_h = (170 - 800)\frac{c_u}{d}$$

Davisson (1970) suggests the more conservative value $k_h = 67 c_u / d$.

The model of k_h linearly increasing with depth may be adopted for NC or slightly OC clays and for cohesionless soils. The values of n_h for cohesionless soils may be evaluated by the expression:

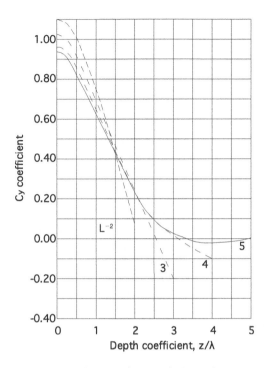

Figure 9.9 Solutions by Matlock and Reese (1956): displacement coefficient for a restrained pile subjected to an horizontal head force.

$$n_h = \frac{A\gamma}{1.35} \tag{9.4}$$

where γ represents the unit weight of the soil; if the soil is submerged, the submerged unit weight γ' will be employed. Suggested values of A and n_h are listed in Table 9.3. For cohesive NC or slightly OC soils some values of n_h are reported in Table 9.4.

9.1.3 Non-linear springs (p–y curves)

The available experimental evidence on the behaviour of laterally loaded piles, as reported in §9.1.1, shows that the load–deflection relationship is markedly non-linear; the conventional linear spring model is thus suited only for rough preliminary evaluation, and requires considerable judgement in the selection of parameter values in relation to the prevailing strain level.

Table 9.3 Suggested value of A and n_h (Eq. 9.4) for cohesionless soil

Relative density	Loose	Medium	Dense
Range of A values	100–300	300–1000	1000–3000
Suggested value of A	200	600	1500
n_h [N/cm^3], dry sand	2.5	7.5	20
n_h [N/cm^3], submerged sand	1.5	5	12

Table 9.4 Suggested values of n_h for cohesive soils

Soil type	n_h [N/cm³]	Source
NC or slightly OC clay	0.2 to 3.5 0.3 to 0.5	Matlock and Reese 1956 Davisson and Prakash 1963
Organic NC clay	0.1 to 1.0 0.1 to 0.8	Peck and Davisson 1962 Davisson 1970
Peat	0.05 0.03 to 0.1	Davisson 1970 Wilson and Hilts 1967
Loess	8 to 10	Bowles 1988

A logical development of the model is the adoption of non-linear spring to represent the soil response. Such an approach has been pioneered by Reese and his co-workers, who developed a series of *p–y* curves for various types of soils on the basis of carefully instrumented field pile tests. Their recommendations may be found in the books of Reese and Van Impe (2001) and Reese *et al.* (2006).

Once the proper *p–y* curves have been selected, they are entered in a computer code solving the problem by an iterative or incremental method; there are a number of commercial codes, such as LPILE developed by Reese and co-workers.

9.1.4 Characteristic load method

Evans and Duncan (1982) and Duncan *et al.* (1994) presented a useful and simple procedure for estimating the load–deflexion behaviour of single piles. They used dimensional analysis to characterize the non-linear behaviour of laterally loaded piles via relationships between dimensionless variables. From the scrutiny of a large number of analyses carried out by the *p–y* curves, they defined a Characteristic Load H_c:

$$\text{for clay: } H_c = 7.34d^2 \left(E_p R_1 \right) \left(\frac{c_u}{E_p R_1} \right)^{0.68}$$

$$\text{for sand: } H_c = 1.57d^2 \left(E_p R_1 \right) \left(\frac{\gamma' d\varphi' k_p}{E_p R_1} \right)^{0.57}$$

and a Characteristic Moment M_c:

$$\text{for clay: } M_c = 3.86d^3 \left(E_p R_1 \right) \left(\frac{c_u}{E_p R_1} \right)^{0.45}$$

$$\text{for sand: } M_c = 1.33d^3 \left(E_p R_1 \right) \left(\frac{\gamma' d\varphi' k_p}{E_p R_1} \right)^{0.40}$$

where d = pile diameter or width; E_p = pile modulus; R_1 = ratio of the moment of inertia of the pile section to that of a solid circular section ($R_1 = 1$ for a solid circular pile); c_u = undrained shear strength of the clay; γ' = submerge unit weight of sand; φ' = effective stress friction angle of the sand (degrees); k_p = Rankine passive pressure coefficient. The authors proposed to characterize the non-linear behaviour of piles through relationships between dimensionless variables, and provided them in tabular form normalizing load and moment to critical load and moment, and displacement to the pile diameter. Later on Brettmann and Duncan (1996) interpolated the numerical results for the lateral load–deflection relationship as follows:

$$\frac{y}{d} = a_h \left(\frac{H}{H_c} \right)^{b_h}$$

while for applied moment loading the relationship is:

$$\frac{y}{d} = a_m \left(\frac{M}{M_c} \right)^{b_m}$$

The relationship between the maximum moment induced in the pile and the applied horizontal load is:

$$\frac{H}{H_c} = a_x \left(\frac{M_{\max}}{M_c} \right)^{b_x}$$

In the above equations: y = groundline deflection; a_h, b_h, a_m, b_m, a_x, b_x = constants; H, M = applied lateral load and moment at the top of the pile. The different constants are given in Table 9.5 as a function of the soil type and the degree of fixity of the pile top.

It is to be pointed out that the characteristic load and moment depend only on the soil strength (c_u, γ', φ') and not on the soil stiffness, making greatly easier the evaluation of soil properties.

The CLM is valid for "long" piles; the minimum length of a pile for the CLM to be valid is listed in Table 9.6. In practice, piles are usually longer than the limit values of Table 9.6.

Further suggestions for the analysis of piles whose top is at depth below ground surface have been given by Ooi *et al.* (2004).

It is evident that the behaviour of a laterally loaded pile is considered not to be affected by the installation technique; as a matter of fact, the values of the

Table 9.5 Constants in the characteristic load method

Constant	Clay		Sand	
	Free head	*Fixed head*	*Free head*	*Fixed head*
a_h	50.0	14.0	119.0	28.8
b_h	1.822	1.846	1.523	1.500
a_m	21.0	–	36.0	–
b_m	1.412	–	1.308	–
a_x	1.22	1.63	0.43	0.66
b_x	0.80	0.87	0.77	0.84

Note
Values of a_x and b_x have been obtained by fitting the numerical values given by Duncan *et al.* (1994), and differ from those reported by Brettmann and Duncan (1996) which appear to be affected by significant errors.

Table 9.6 Minimum pile length for the validity of CLM

Soil type	$\dfrac{E_p R_1}{c_u}$	$\dfrac{E_p R_1}{\gamma' d\varphi' k_p}$	Minimum pile length $\dfrac{L}{d}$
Clay	100 000		6
Clay	300 000		10
Clay	1 000 000		14
Clay	3 000 000		18
Sand		10 000	8
Sand		40 000	11
Sand		200 000	14

parameters needed to apply the method are given in Table 9.5 as a function of the soil type and degree of fixity, with no reference to the installation techniques.

This assumption may be checked by back-analysing the available lateral load tests to deduce the values of the parameters taking into account the installation technique. A database of some 20 sufficiently well-documented case histories collected by Landi (2006) has shown a definite dependence of a_h on undrained strength c_u for piles in clay, irrespective of the installation technique (Figure 9.10); the experimental data are well represented by the equation:

$$a_h = 1.1\left(\frac{c_u}{p_r}\right)^{0.75}$$

where $p_r = 1\,\text{kPa}$ is a reference pressure. The Brettman and Duncan value $a_h = 50$ corresponds to a rather stiff clay ($c_u = 170\,\text{kPa}$), while smaller values are obtained for softer clays.

For bored piles in sand the available data show again a dependence of a_h on the friction angle φ' (Figure 9.11) that can be expressed by the equation:

$$a_h = 4.92\varphi' - 83$$

where φ' is again expressed in degrees. The Brettman and Duncan value $a_h = 119$ corresponds to a value of $\varphi' = 41°$, while smaller values are obtained for looser sands.

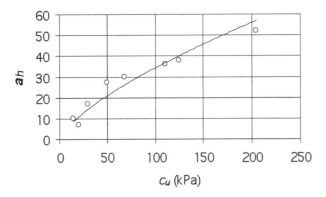

Figure 9.10 Dependence of a_h on c_u for piles in clay.

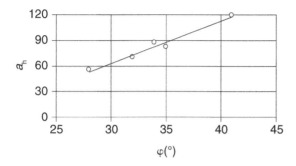

Figure 9.11 Dependence of a_h on φ' for bored piles in sand.

The values of H_c and M_c increase with increasing c_u or φ'; the net effect of the increase of a_h, H_c and M_c, as it was to be expected, is however a decrease of the deflection as the soil becomes stiffer or denser.

For driven piles in sand a reliable relationship between a_h and φ' has not been found, due to the data scatter. The available values of a_h fall in the range 40–120, with a tendency to decrease with increasing φ'.

The values of parameter b_h given by Brettman and Duncan have been substantially confirmed by our back-analyses.

The parameters a_m, b_m, which appears in the moment-displacement equations have not been investigated for the substantial lack of experimental data. The relationship between the applied lateral load and maximum bending moment will be discussed in the next sections. However, it can be seen from Table 9.5 that the values of the exponent b_x are closer to unity than b_h. This is in agreement with the experimental evidence that the lateral load–maximum moment relationship is nearly linear.

The CLM is a useful and practical tool, permitting a prediction of the non-linear load–deflection relationship without a complex characterization of the non-linear deformation characteristics of the soil. As a matter of fact, the only soil parameters needed are the strength parameters c_u or φ' and the unit weight γ.

9.1.5 Boundary Element Method

Early application of the Boundary Element Method (BEM) to vertical piles under horizontal or moment loading at the pile head can be traced back to Douglas and Davis (1964) and Spillers and Stoll (1964), who first proposed a simple way to deal with soil non-linearity by introducing a limiting value to the soil–pile interaction force. Poulos (1971a) modelled the pile as a linearly elastic strip interacting with a linearly elastic homogeneous and isotropic half space, and used the solutions obtained by Douglas and Davis (1964) for the horizontal displacement induced by a horizontal load uniformly distributed over a rectangular area to develop the first systematic BEM parametric study of the problem. Later on other studies have considered factors as subsoil models other than the homogenous half space and non-linearity of the flexural behaviour of pile shaft and of the pile–soil interaction. A broad synthesis of these investigations is reported in the treatise by Poulos and Davis (1980).

A short description of the method, as originally developed by Poulos (1971a), is reported. The pile–soil interface is discretized into n rectangular areas of width d and length $l = L/n$; the distribution of stress acting at the interfaces is approximated (Figure 9.12) by n values of horizontal normal stress p uniformly distributed with constant intensity over each area.

As in §5.2.4 for piles under vertical load, the problem is solved by equating the horizontal displacement of the pile to those of the half space, both calculated at mid height of each element. Two further equations are obtained by imposing equilibrium between the external forces (horizontal load H and moment M) and the pressure distribution along the pile shaft.

The horizontal displacement at mid-height of the pile element i may be written as:

$$y_i = -\sum_{j=1}^{n} a_{ij} P_j + y_0 + \theta_0 \cdot z_i$$

with:

$$P_j = p_j dl$$

$$a_{ij} = \frac{z_i^3}{3 E_p I_p} + \frac{z_i^2 (z_j - z_i)}{2 E_p I_p} \qquad \text{if } z_i \le z_j$$

$$a_{ij} = \frac{z_j^3}{3 E_p I_p} + \frac{z_j^2 (z_i - z_j)}{2 E_p I_p} \qquad \text{if } z_i > z_j$$

and y_0, θ_0 representing the unknown horizontal displacement and rotation at the pile head. In such a way the displacements of the points belonging to the pile are expressed as linear functions of the $(n+2)$ unknowns p_j, y_0 and θ_0.

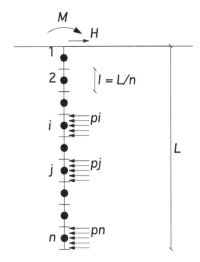

Figure 9.12 BEM scheme for a pile under horizontal load H and moment M (after Poulos 1971a).

The horizontal displacement w_{ij} induced at a point i belonging to the half space by a horizontal force P_j applied in the point j can be obtained by the Mindlin (1936) solution:

$$w_i = \sum_{j=1}^{n} w_{ij} = \sum_{j=1}^{n} b_{ij} P_j$$

where

$$b_{ij} = \frac{A_{ij} + B_{ij} + C_{ij}}{16 \, \pi \, G(1-v)}$$

and

$$G = \frac{E_s}{2(1+v_s)}$$

$$A_{ij} = \frac{(3-4v_s)}{R_{1ij}} + \frac{1}{R_{2ij}} + \frac{x_{ij}^2}{R_{1ij}^3} + \frac{(3-4v_s)x_{ij}^2}{R_{2ij}^3}$$

$$B_{ij} = \frac{2 c_j \, z_i}{R_{2ij}^3}\left(1 - \frac{3x_{ij}^2}{R_{2ij}^2}\right)$$

$$C_{ij} = \frac{4(1-v_s)(1-2v_s)}{R_{2ij} + z_i + c_j}\left(1 - \frac{x_{ij}^2}{R_{2ij}\left(R_{2ij} + c_j + z_i\right)}\right)$$

$$r_{ij} = \sqrt{x_{ij}^2 + y_{ij}^2}$$

$$R_{1ij} = \sqrt{r_{ij}^2 + \left(z_i - c_j\right)^2}$$

$$R_{2ij} = \sqrt{r_{ij}^2 + \left(z_i + c_j\right)^2}$$

E_s, v_s are Young modulus and Poisson ratio of the elastic half space. The meaning of the symbols is illustrated in Figure 9.13. The unknown soil displacements are again linear functions of the unknown forces P_j.

As stated above, the compatibility is imposed equating the displacement of pile and soil in the n points:

$$y_i = w_i \quad i = 1 \text{ to } n \tag{9.5}$$

The two additional equilibrium equations are written:

$$\sum_{j=1}^{n} P_j = H$$

$$\tag{9.6}$$

$$\sum_{j=1}^{n} P_j \cdot z_j = -M$$

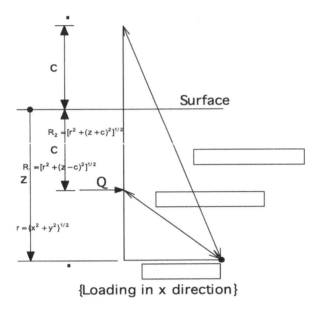

Figure 9.13 Mindlin problem for horizontal load in elastic half space.

The system of $n+2$ linear equations made by Eq. 9.5, and Eq. 9.6 may be solved for the $n+2$ unknowns P_j, y_0 and θ_0.

If the pile head is not free to rotate the second equilibrium equation is omitted and the rotation θ_0 is set to 0.

After solving the system, the displacement and the rotation at the pile head may be expressed as:

$$y_0 = \frac{H}{E_s L} \cdot I_{yH} + \frac{M}{E_s L^2} \cdot I_{yM}$$

$$\theta_0 = \frac{H}{E_s L^2} \cdot I_{\theta H} + \frac{M}{E_s L^3} \cdot I_{\theta M}$$

If the pile head is not free to rotate the displacement at the pile head may be expressed as:

$$y_0 = \frac{H}{E_s L} \cdot I_{yF}$$

Poulos and Davis (1980) report exhaustive plots of the various influence factors I as a function of the dimensionless quantities (L/D, v_s, $K_r = E_p I_p/E_s L^4$). An example is provided in Figure 9.14.

Kuhlemeyer (1979) and later on Randolph (1981) observed that in a pile under horizontal load the curvature, bending moment and shear vanish within a depth of around ten diameters from the pile head, generally much smaller than the actual pile length. For this reason an effective or critical length L_c should be considered, rather than the actual length L of the pile. According to Randolph (1981) the critical length of a pile embedded in an homogeneous half space may be defined as:

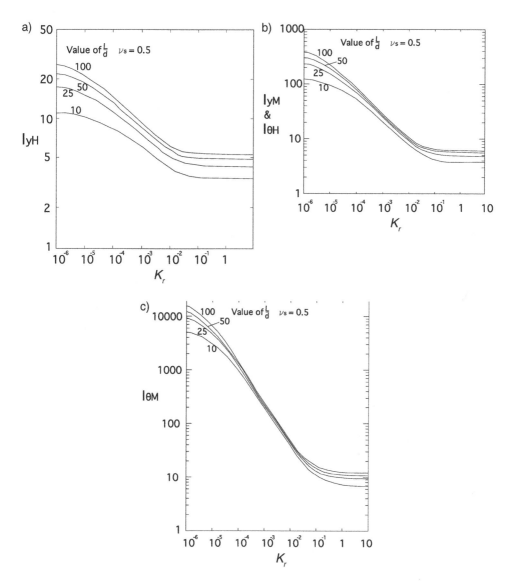

Figure 9.14 Influence coefficient for piles under horizontal load and moment (source: Poulos and Davis 1980).

$$\frac{L_c}{r} = 2 \cdot \left[\frac{E_p}{G \cdot \left(1 + \dfrac{3v}{4}\right)} \right]^{2/7}$$

Randolph (1981) also provided simple approximate expressions of the displacement and rotation of the pile head under horizontal and moment loading.

There are a number of computer programs available for the analysis of a single pile subjected to horizontal and moment loading, based on the elastic continuum model solved by BEM. The program STHOP (Abagnara 2009; Abagnara and Russo 2007), representing an evolution of the program NAPHOL (Landi and Russo 2005), relies upon the following assumptions:

- horizontally layered elastic soil. When computing the displacement w_{ij} the Mindlin solution is applied characterizing the layers crossed by the pile shaft by the average of the Young's moduli between the point i and j (Poulos 1973);
- pile with constant or stepwise variable section;
- non-linear behaviour of the reinforced concrete pile section;
- limiting pressure at the pile–soil interface.

The limiting pressure in the program STHOP is an input datum and different values for each pile segment corresponding to different soil layers may be selected.

The authors carried out an extensive validation of the computer code against available experimental evidence and reached the following conclusion:

- the best option for piles in clay is to evaluate the limiting soil reaction as $p = 9c_u d$ (Broms 1964a) starting from a depth $z = 2d$ downwards; from ground level ($z = 0$) to $z = 2d$ the limiting soil reaction can be linearly interpolated between $p = 0$ at ground level and $p = 9c_u d$ at $z = 2d$;
- for piles in sand the limiting soil reaction may be set to $p = k_p^2 \gamma dz$ (Barton 1984); a slight improvement may be obtained reducing the limiting soil reaction from ground level ($z = 0$) to $z = d$ to $p = k_p \gamma dz$.

The very detailed indication is justified by the significance of soil reaction at shallow depth.

The elastic modulus in the model, which has large influence only on the very initial part of the load–displacement relationship, is fixed according to the following suggestion:

- for piles in clay $E_u = 1000–1200\,c_u$
- for piles in sand $E/(\gamma d) = 150\,\varphi° – 2300$

In order to evaluate the capability of the simple elastic-perfectly plastic bilinear model to reproduce the actual non-linear behaviour of the pile, some comparisons with available experimental results are examined.

Reese *et al.* (1975) report the results of a horizontal load test on a driven steel tubular pile with an external diameter $d = 64$ mm and a total length $L = 15.2$ m, in stiff overconsolidated clay. In Figure 9.15 the load–displacement curve of the pile head is reported. Code STHOP has been applied according to the previous suggestions for piles in clayey soils and using c_u values increasing with the depth starting from 25 kPa (at ground level) to 1100 kPa at a depth of 10 m as suggested by Reese *et al.* (1975).

The agreement with the experimental results is remarkable.

Another horizontal load test on a driven steel tubular pile with an external diameter $d = 273$ mm and a total length $L = 13.1$ m, again in overconsolidated clay, is reported in Figure 9.16 (Brown *et al.* 1987, after Reese and Van Impe 2001). Code STHOP has been applied according to the previous suggestions for piles in clayey

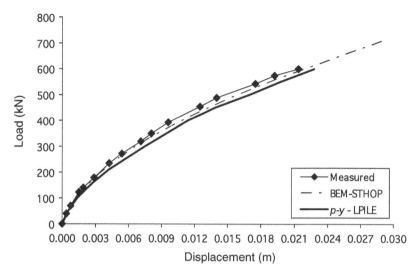

Figure 9.15 Comparison between computed results and experimental results (data from Reese *et al.* 1975).

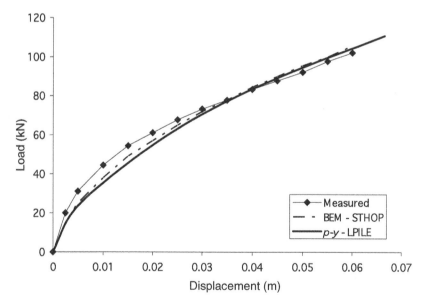

Figure 9.16 Comparison between computed results and experimental results (data from Brown *et al.* 1987).

soils and using c_u values increasing from 54 kPa (at ground level) to 148 kPa at a depth of 5.5 m as suggested by the authors; the agreement is again very satisfactory.

It may be observed that Reese and Van Impe (2001) have used the same two case histories to test the computer program LPILE, based on the transfer curve approach; Figure 9.17 reports the shape of the *p–y* curve they adopted, typical of overconsolidated clays. The results of their fitting are also reported in Figures 9.15 and 9.16.

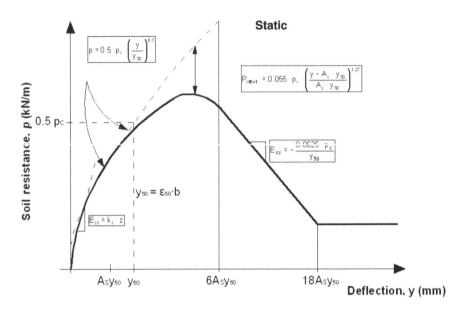

Figure 9.17 Suggested *p–y* curve for stiff overconsolidated clay (source: Reese and Van Impe 2001).

It can be seen that the simple elastic approach with a limiting pressure at the pile soil-interface implemented into the BEM code is capable of providing the same excellent fit of the observed behaviour as that obtained by the more complex *p–y* curve approach. Such a finding, confirmed by the back-analyses of quite a number of case histories, definitely shows that the simple bilinear approach to soil non-linearity, usually implemented in BEM codes, is sufficient to get a fairly accurate prediction of the non-linear load–displacement relationship without the need of adopting more complex *p–y* functions.

9.1.6 Maximum bending moment

The non-linear BEM code described in the previous section is capable also of good predictions of the bending moments along the pile shaft. A satisfactory prediction of the relationship between horizontal load and maximum moment along the pile shaft relies upon the appropriate modelling of the non-linear behaviour not only of the soil but also of the structural material, at least for reinforced concrete sections.

As discussed in §9.1.1 the load–moment relationship is nearly linear up to high load level. In Figure 9.18 an example of the kind of results that are obtained by BEM is reported. The examined case is that of an elastic pile subjected to increasing lateral load with the subsoil modelled as an elastic half space (E_s, v_s); a limiting value of the pressure between the pile and the soil is imposed. In Figure 9.18 the limit pressure has been evaluated as $9c_u$, with c_u linearly increasing with depth; the validity of the following remarks, however, is not limited to this specific case.

The load–deflection curve obtained by BEM is markedly non-linear, while the load–maximum moment curve is only slightly curved with the concavity upwards.

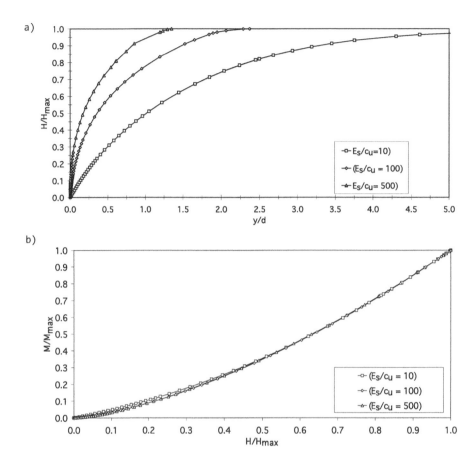

Figure 9.18 Example of BEM applications (code STHOP) for an elastic pile embedded in a clayey soil.

A more realistic non-linear model for the pile body would increase the non-linearity of the load–settlement relationship, while further attenuating the slight curvature of the load–maximum moment relationship.

It may be seen that the load–deflection curve is strongly affected by the deformation characteristic of the soil (values of E_s/c_u), while the lateral load–maximum moment relationship is nearly linear and practically unaffected by the relative stiffness of the pile–soil system.

The reason of this unexpected behaviour is clarified by the results of BEM, plotted in Figure 9.19. For any value of the applied load, the mobilized pressure at the pile–soil interface attains its limiting value down to about the depth where maximum moment occurs. A sort of progressive downward failure mechanism at the pile soil interface occurs, such as shown by the results of Figure 9.19. This mechanism, and hence the maximum bending moment, is not significantly affected by the relative stiffness of the pile.

Direct experimental evidence of the above mechanism cannot be provided because of the lack of measurements of the lateral pressure exerted by the soil on the pile;

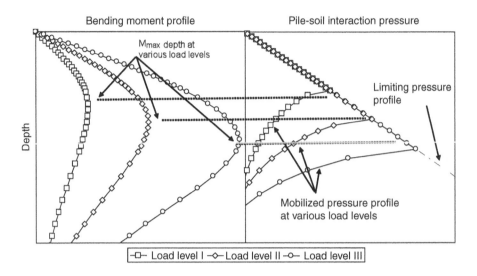

Figure 9.19 Typical BEM results (code STHOP): bending moments and mobilized lateral soil
pressure versus depth.

some indications, however, may be obtained by double derivations of the bending
moments profiles. The process is far from straightforward and the results are
strongly influenced by the details of smoothing and integration; some results, however, are reported in Figure 9.20.

The above findings suggest a method to predict the load–maximum moment relationship, much simpler than either $p–y$ curves or BEM.

At a given load, if the limiting pressure profile is assumed to act in the upper part
of the pile, the horizontal equilibrium allows the determination of the depth at which

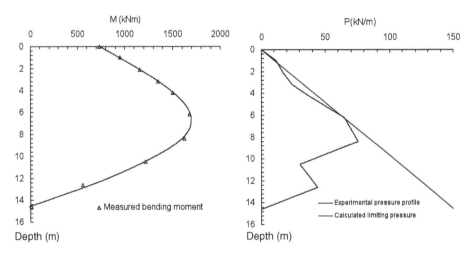

Figure 9.20 Experimental results for bending moments and mobilized lateral soil pressure
versus depth (centrifuge test on a square pile $d = 12$ mm – NC clay, Ilyas *et al.*
2004).

the shear in the pile is equal to zero. At the same depth the maximum bending moment can be determined by simple statics.

This method was tested on some 20 case histories, extracted from a database reported by Landi (2006), where the load–maximum moment curve was available.

The cases cover a broad and varied population: steel and reinforced concrete piles, driven or bored, in soft or stiff clays and in loose or dense sand. Table 9.7 summarizes the average difference between the measured values and the computed ones at three different displacement levels. A sort of initial or low displacement comparison was set at $y=0.02d$, while the remaining two comparisons refer to 50% and 100% of the final displacement attained in the test. The agreement between computed and observed values is satisfactory.

The success of the method relies on a detailed characterization of the strength properties of the upper soil layers; typically the maximum depth involved in the calculation is within 6–$8d$.

9.1.7 Finite Element Method

The horizontal displacement and the bending moment along a pile subjected to a horizontal load may be analysed by finite elements. Because of the loading condition it is not possible to take advantage of the axial symmetry of the pile body; the problem to be analysed is thus three-dimensional and requires large computational resources. This is probably the reason why the published studies relating to horizontally loaded piles are even less than those dealing with piles under vertical load.

In the late 1970s some approximate 3D solutions starting from 2D FEM models and making use of Fourier's series have been published (Kuhlemeyer 1979; Randolph 1981; Krishnan *et al.* 1983). Later on Brown and Shie (1990a, 1990b) published a full 3D FEM study of a single pile or a row of piles embedded into the soil modelled as an elastic-perfectly plastic material and subjected to horizontal load. The study was aimed to derive simplified p–y curves for both the single pile and the same pile in a row. Yang and Jeremic (2002, 2003) extended the analyses to small groups of piles and tried also to outline the influence of the interaction among the piles on the moments along the pile shaft.

At present, the use of FEM in design seems to be still out of question, because of the 3D nature of the problem and the difficulties connected with a proper characterization of the soils. The value of FEM analyses, however, may be found in the development of benchmark solutions for the assessment of the simplified methods presented above.

Table 9.7 Accuracy of the proposed method for the calculation of the load–moment curve

	Clay			Sand		
	$y=0.02d$	$y=50\%\ y_f$	$y=100\%\ y_f$	$y=0.02d$	$y<h=50\%\ y_f$	$y=100\%\ y_f$
Error (%)	16	13	8	15	12	10

9.2 Pile groups

9.2.1 *Experimental evidence*

Much like piles under vertical loads, the response of a laterally loaded pile group with relatively closely spaced piles is quite different from that of a single pile. The differences are mainly due to the interaction between piles through the surrounding soil, the rotational restraint exerted by the cap connecting the piles at the head, the additional resistance to lateral load provided by frictional resistance at the cap–soil interface, and passive resistance on the cap if totally or partially embedded.

The experimental evidence is rather scanty, compared to that available for vertical load, and is mainly related to the first two items. Relatively small groups have been tested, typically small scale models and full scale foundations with 2 to 16 piles. In recent years the use of centrifuge allowed the study of larger groups (16 to 21 piles).

A comprehensive review of the experimental evidence is reported by Mokwa (1999). Later on, valuable experimental investigations have been carried out by Borel (2001); Rollins and Sparks (2002); Ilyas *et al.* (2004); Rollins *et al.* (2005).

Calling the trailing row the first row of the group and the leading row the last one (Figure 9.21), centrifuge (Barton 1984) and full scale tests (Ochoa and O'Neill 1989) revealed that, for a given horizontal load parallel to the columns of a group, the piles in the leading row carried more load than those in the trailing one, even at load levels far from failure.

Selby and Poulos (1984) measured shears and bending moments in the leading piles larger than that in the central and trailing piles in 1 G model tests; they called this effect *shielding*. Brown *et al.* (1988) observed the same effect in a full-scale test on a 3^2 pile group and introduced the term *shadowing* to mean the phenomenon for which the soil resistance of a pile in a trailing row is reduced because of the presence of the leading pile ahead of it.

A rather large number of experiments have been carried out in the last decade with the aim of deriving "general" rules to adapt p–y curves to account for group effects. Brown *et al.* (1988) introduced the concept of p–y multiplier f_m, a multiplier of the p values capable of stretching the p–y curve for the single pile to account for the interaction among the piles in a group (Figure 9.22). The value of f_m is in the

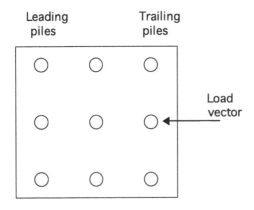

Figure 9.21 Leading and trailing piles for a given load vector.

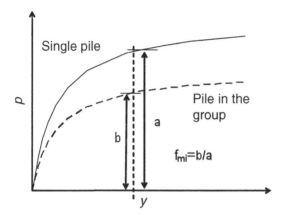

Figure 9.22 p–y multipliers for group effects.

range zero to one. A major part of the latest experimental work has been devoted to the determination of the *p–y* multipliers. The latest experiments are at field scale or in centrifuge; the use of 1 G small-scale tests has been abandoned having recognized how misleading the obtained results could be in terms of stiffness, because of scale effects.

The collected observations (Brown *et al.* 1988; McVay *et al.* 1998; Rollins *et al.* 1998; Mokwa 1999) suggest that:

- the multipliers can be defined for rows orthogonal to the direction of the load vector, being the differences among the piles in a row almost negligible;
- after the third leading row the same multiplier applies to the other rows, except the trailing one;
- the multipliers are independent of the soil type, pile type and load level, and depend essentially on the spacing;
- at a spacing above six to eight diameters in the direction of the load vector, and four diameters in the orthogonal direction, the interaction among piles is negligible and the multipliers can be assumed equal to one.

A parameter frequently used to compare the response of single pile and pile groups under horizontal load is the group efficiency:

$$G_e = \frac{\dfrac{H}{n}}{H_S} \tag{9.7}$$

where H is the total horizontal load producing a given displacement, n the number of piles in the group and H_S the horizontal load carried by a single pile at the same horizontal displacement. It is worth noting that, when dealing with piles under vertical load, the group effect is usually expressed through a multiplier of the settlement at a given load; on the contrary for piles under horizontal load, a reduction of the load per pile at a given displacement is often used. This reflects the fact that in the former case the emphasis is on displacements, in the latter the main design issue is the stress in piles.

The efficiency G_e defined by Eq. 9.7 can be easily expressed in terms of the p–y multipliers as follows:

$$G_e = \frac{\sum\limits_{i=1}^{m} f_{mi}}{m} \qquad (9.8)$$

where f_{mi} is the multiplier of the i-th row while m is the number of rows in the group. Assuming constant values for the p–y multipliers, irrespective of the load or the displacement level, the efficiency of the group G_e is also constant.

In Figures 9.23 and 9.24 some experimental values of the efficiency G_e are plotted against the displacement normalized by the diameter of the pile. The data reported in Figure 9.23 were obtained by field tests while those in Figure 9.24 by centrifuge tests. In both cases the experiments were carried out under free-head conditions for both the single piles and the pile groups.

For the data reported in Figure 9.25, obtained also by centrifuge tests, the single pile was tested under free-head conditions while the pile group had a rotational restraint at the pile head. All the data reported in Figures 9.23, 9.24 and 9.25 refer to pile groups with a constant spacing $s = 3d$.

The efficiency G_e is always below unity and decreases with increasing displacement (Figures 9.23 and 9.24). In Figure 9.25 the efficiency is above unity, as expected, due to the rotational restraint at the head of the piles in group, but again G_e significantly decreases with increasing displacement. Figures 9.24 and 9.25 show a dependence of the efficiency G_e on the size of the group: the larger the group size, the lower the efficiency.

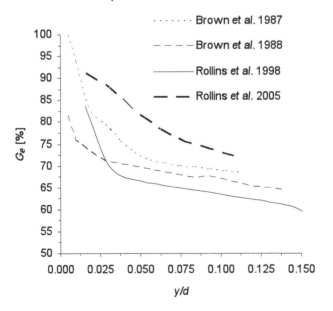

Figure 9.23 Efficiency G_e vs. displacement for field tests of small pile groups under horizontal load.

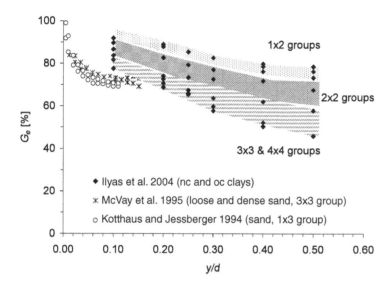

Figure 9.24 Efficiency G_e vs. displacement for pile groups (free head) of different size under horizontal load (centrifuge tests).

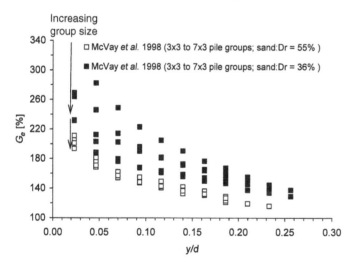

Figure 9.25 Efficiency G_e vs. displacement for pile groups (fixed head) of different size under horizontal load (centrifuge tests).

The tests reported in Figure 9.25 were carried out just increasing the number of rows and keeping constant the number of piles in each row because of widespread belief that the p–y multipliers can be assumed constant for each row and independent of the number of piles contained in the rows.

To compare the behaviour of pile groups under horizontal and vertical load, the definition of efficiency G_e can be easily extended to pile groups under vertical load, just exchanging the shear H with the vertical force Q. In Figure 9.26 the data provided by three field tests on small pile groups and two large pile groups under

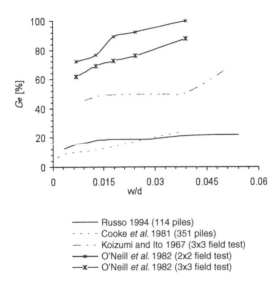

Russo 1994 (114 piles)
Cooke *et al.* 1981 (351 piles)
Koizumi and Ito 1967 (3x3 field test)
O'Neill *et al.* 1982 (2x2 field test)
O'Neill *et al.* 1982 (3x3 field test)

Figure 9.26 Efficiency G_e vs. displacement for pile groups under vertical load.

vertical loads are reported. As in the case of horizontal load, the efficiency under vertical load decreases with increasing the size of the group. The effect of the displacement, on the contrary, is the opposite; the efficiency increases with increasing displacement. The interaction mechanisms under vertical and horizontal load are thus very different.

Coming back to pile under horizontal load, the practice of fixing a unique multiplier for each row is not so obvious; a summary of the available data on the load sharing among the piles in a group will be used to clarify the point. In all the experiments uniform settlement was imposed on the group and consequently the load is not uniformly distributed on piles.

Morrison and Reese (1986) carried out a field test on a 3^2 pile group in sand and reported a maximum difference between pairs of adjacent piles belonging to the same rows of about 33%, while for pairs belonging to the same column the difference was slightly above 100%.

The field test of a 4^2 pile group in sand carried out by Ruesta and Townsend (1997) revealed differences above 100% between piles in the same row or in the same column. McVay *et al.* (1998) performed centrifuge tests on groups of variable size and found differences between two adjacent piles in the same rows not always negligible and, sometimes, comparable to the differences between two adjacent rows. Similar results are reported by Ilyas *et al.* (2004).

Rollins *et al.* (2005) published the results of a field test on a 3^2 group at spacing of 3.3d. Substantial differences in the load sharing among piles in the same row were observed. The internal piles carried systematically the lowest load. The ratio between the centre and the outer pile loads in the same row is in the range 65% to 80%; the same range applies also for piles belonging to different rows. Even if the lower interactivity among piles placed orthogonal to the direction of the load vector compared to that among piles aligned in the direction of the vector is widely accepted evidence, the above data show that the effects in a group are not at all negligible.

The bending moment in the piles being more critical in the design than the horizontal displacement, it is also interesting to summarize the available experimental evidence on this item. The experimental results are affected by some scatter, probably due to the experimental difficulties and also to the detail of the rotational restraint imposed at the pile head. Some general trends however can be identified.

Data collected for the cases where both the single pile and the pile group were tested with free head conditions are reported in Figure 9.27. The ratio between the maximum moment in different piles in a group and that in a single isolated pile is plotted against the displacement of the pile group; the ratio is evaluated at the same average load per pile. The data are rather scattered but the values of the moment in the piles belonging to the group are generally larger than those in the single pile, the increase being a growing function of the displacement. Larger moments occur for the piles belonging to the Leading Row (LR) if compared either to the middle (MR) or the trailing one (TR). These findings show, for instance, that it would be unsafe to evaluate the moment in a pile within a group as the moment of a single pile subjected to the average load even if this is commonly done in engineering practice.

This occurs for two reasons. The maximum bending moment in a pile subjected to a horizontal load at the head depends on: (i) the intensity of the load and (ii) the depth needed for the pile to develop a sufficient reaction into the surrounding soil; the deeper the soil reaction the higher the bending moment with the same head load. In a pile group the applied load is not uniformly shared among the piles, the load distribution being a function of their position within the group. The higher moments of the leading piles in Figure 9.27 are thus partially due to head loads higher than average. The interaction among the piles, furthermore, develops a deeper reaction in the surrounding soil, compared to that of a single isolated pile. This trend is more

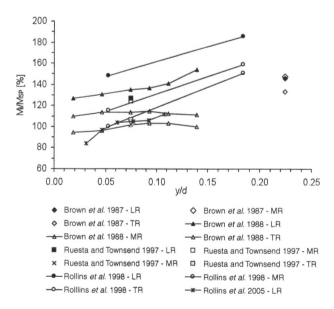

Figure 9.27 Ratio between bending moments of piles in a group and those in a single isolated pile vs. displacement.

marked for the trailing piles than for the leading ones. This may well explain why in some cases differences in the observed maximum bending moments between leading and trailing piles are not very large, even when the head load on the leading piles is much higher than average.

9.2.2 Methods of analysis

Interaction factors method

Poulos (1971b) first proposed the interaction factors method to analyse pile groups under horizontal load, using the theory of elasticity to derive the interaction factor α_{ij} between a loaded pile i and an unloaded adjacent pile j. The matrix of interaction factors is symmetric and independent of the load level. Compared to the case of vertical load where α_{ij} is only a function of the spacing of the piles i and j, computational complexity is slightly increased by the dependence of the interaction factor on the angle β between the load vector and the line connecting the pile i and j.

In Figure 9.28 (Poulos 1971c) an example of the variation of α_{ij} with the angle β is reported; it attains two symmetric maximums at $\beta=0°$ and $180°$ and a minimum at $\beta=90°$. Poulos and Davis (1980) presented a wide range of charts of interaction factors; Randolph and Poulos (1982) developed analytical formulae based on the concept of critical length, LC. The method has not gained the same popularity as for the analysis of pile groups under vertical loads. Some relatively widespread computer programs as DEFPIG (Poulos 1988) or PIGLET (Randolph 1987) are based on the interaction factors but the available evidence to demonstrate their accuracy or to define their limits is far from exhaustive.

The experimental evidence shows clearly that:

a   the matrix of the interaction factors is non symmetric (Ochoa and O'Neill 1989; Schmidt 1981); the value of α at $\beta=180°$ may be as low as 0.5 times the value of α at $\beta=0°$ at low load level.
b   the interaction factors are not independent of the load level (Ochoa and O'Neill 1989); they tend to increase with increasing load level.

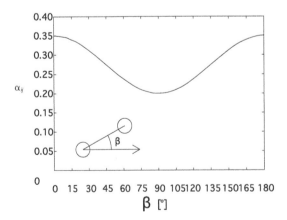

Figure 9.28 Typical variation of interaction factor between a couple of piles α_{ij} as a function of the angle β.

The experimental evidence reported in the previous paragraphs, showing the decreasing efficiency of the group of piles with increasing settlement, can be easily used to further support this last item.

It appears that an accurate prediction of the behaviour of pile groups under horizontal load should take into proper account the points (a) and (b) above, with interaction factors function of the angle β and of the load level on each pile of the group. The method would then lose its simplicity, which is probably the main reason of its success for the analysis of pile groups under vertical load.

Non-linear p–y curves

Along a different research path, the spring model introduced by Matlock and Reese (1956, 1960) has been extended in subsequent years to the analysis of pile groups. The p–y method cannot actually account for the interaction through a continuum both along the single pile and among piles in a group. On the other hand it easily allows the consideration of soil properties changing with the depth along the shaft of the pile.

To analyse a pile group, the shape of the p–y curves have to be adapted or, better, stretched to simulate the interaction effect among the piles of the group through the soil. The adaptation is typically based on the p–y multipliers f_m introduced in §9.2.1 and is far from straightforward. It should take into account at least the main factors influencing the interaction between the piles of a group, such as:

a   the mechanical properties of the soil and, above all, the stiffness variation of the soil layers with depth;
b   the geometry of the pile, the geometry of the pile group, the total number of the piles, the load level on each pile.

The influence of such factors can be deduced on a theoretical basis and is confirmed by the experimental evidence collected and presented in §9.2.1.

It is clear that to account for the influence of all these factors in a unique parameter as the p–y multiplier is difficult and largely relies upon engineering judgement and experience.

The experimental findings presented in §9.2.1 stimulate some further comments:

- The widespread assumption that p–y multipliers f_m are independent of displacement appears a reasonable choice only if a particular displacement level is of concern, i.e. an equivalent secant approach is adopted. On the contrary, if the objective of the analysis is an accurate prediction of the full load–displacement curve, the multipliers should be better considered as decreasing functions of the displacement level.
- The p–y curve method for the single pile and its extension to the pile group is an empirical procedure; both the shape of the curves and their multipliers can only be deduced by ad hoc experiments. Until observations on large size groups are available, the reliability of the method cannot be assessed. Bearing in mind the marked influence of the size of the groups on the efficiency under vertical loads, and the lack of experimental data on large pile group subjected to horizontal load, the use of the multipliers obtained by experiments on small groups in the analysis of large pile groups under horizontal load is questionable and could be overly unconservative.

Boundary Element Method

The Boundary Element Method is one of the options for analysing the behaviour of pile groups under horizontal loading. The description of the method for a single vertical pile under horizontal loading is reported in §9.1.5.

The extension to the analysis of a pile group is based on the adoption of a Green function to calculate the interaction through the elastic continuum used to model the subsoil between each couple of piles' segments even when belonging to different piles in the group. The function proposed by Mindlin (1936) already described in the section dedicated to the single pile is usually adopted. In the case of a pile group every pile is divided into n segments and for this reason the computational resources to solve even the linear elastic problem is huge. For a group with m piles a fully populated square matrix with $(n \times m)^2$ terms must be assembled and handled to solve the system of linear equations.

Poulos (1971b, 1971c), Poulos and Davis (1980) and El Sharnouby and Novak (1986) have first applied the method either to the simplest group made by a couple of piles or to a generic group of m piles embedded in an elastic half space.

Several studies have been published in the same period to extend the application of the BEM to an elastic medium with Young's modulus increasing with the depth (Banerjee and Davies 1978) or to introduce non-linearity of soil behaviour (Davies and Budhu 1986). Usually these studies were limited to the single pile, for the limited computational resources available at that time.

No significant further contributions on this topic may be found in the subsequent years.

In §9.1.5 the computer program STHOP based on BEM for the analysis of a single pile under horizontal loading has been presented. The same program is capable of analysing pile groups under horizontal loading and bending moments with a constraint at the pile head which can vary between the two extremes described by a hinge or by a perfectly fixed head.

In the analysis of pile groups the program STHOP adopts the following further assumptions:

a   non-symmetrical extinction distance for the interaction between a couple of segments belonging to different piles in the group; this distance is defined according to the suggestion by Reese and Van Impe (2001) and is sketched in Figure 9.29;

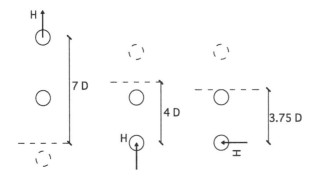

Figure 9.29 Extinction distance for interaction between couple of piles (after Reese and Van Impe 2001).

b   check of block failure for small spacings; the program requires a limiting pressure profile as defined for the single pile but when the spacing between the piles
is so small as not to allow the full development of soil strength a reduced profile
is adopted according to the mechanism of the block failure. The so-called *shadowing* effect, described in the §9.2.1, is thus implemented in the software.

To validate the program with its main assumptions and the method for the determination of the required input parameters a total of 36 case histories have been collected and a prediction exercise has been carried out. Most of the case histories
derive by centrifuge tests and a few from full-scale tests; furthermore, the number of
piles in the group is generally rather small. The largest field test is on a 15 piles
group, while the largest group tested in the centrifuge includes 21 piles.
   In all the examined cases, a load test on a single pile was also available.
   The program was applied both on the single pile and on the pile groups. Some judgement was applied in the simplification of the soil profile to predict the behaviour of the
single pile. The same model was then used to predict the behaviour of the pile group.
   In Figures 9.30 and 9.31 the comparison between experimental and computed
results is illustrated. The computed and observed values of the average load per pile

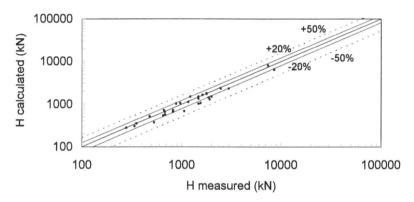

Figure 9.30 Comparison between computed and measured load at a displacement level
$y = 0.5 y_{max}$.

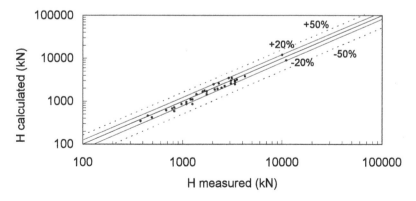

Figure 9.31 Comparison between computed and measured load at a displacement level
$y = y_{max}$.

at a given displacement are compared. The main aim was to test the capability of the method to predict the available experimental full load–displacement curves; for such a reason Figure 9.30 refers to a displacement equal to half of the maximum value reached during the test, while Figure 9.31 refers to the maximum displacement.

It can be seen that the agreement is rather satisfactory with most of the cases falling in the range ±20% and all the cases falling in the range ±50%.

Part IV
Analysis and design of piled rafts

10 Piled rafts

10.1 Introduction

Conventional design procedures for piled foundations, as presented in the previous chapters of this book, often result in an overly conservative design. Among the many factors contributing to such a result, the following ones are probably the most significant:

1   the bearing capacity of single pile is generally evaluated by methods heavily relying on empiricism, although based on the theoretical concepts of soil mechanics. The databases of measured bearing capacities are generally affected by large scatter, and the proposed design methods unavoidably tend to fit the lower bound of experimental values;
2   in evaluating the bearing capacity of pile groups from that of single pile, group effects implying efficiency values much lower than unity are assumed, on the basis of conservative and often outdated literature suggestions;
3   the contribution to the response of the foundation of the structural element (footing, beam, raft), connecting the head of the piles and in contact with the ground, is neglected (in fact, codes and regulations often inhibit its consideration).

No substantial improvement of the practice under item 1 above can be foreseen at present, and again one cannot but agree with Poulos *et al.* (2001) that it is very difficult to recommend any single approach to the evaluation of the bearing capacity as being the more appropriate. Given the very nature of the problem, further development of regional design methods combining local experience of both piling contractors and designers appears the only viable perspective. In any case, a precise evaluation of the bearing capacity fortunately tends to lose importance in a settlement based design, as outlined below.

Concerning the item 2, efficiency values lower than unity depend on a combination of two main factors: (i) a relatively small spacing among the piles (three diameters or less), and (ii) poor constructional techniques with either reduction of the density of granular soils or remoulding and destructuration of cohesive soils.

The trend toward a smaller number of piles at spacing even only slightly larger than usual (4–5 diameters) is going to substantially reduce the influence of the former factor, while the adoption of adequate installation techniques and tools is working against the second.

The most important factor, however, is the contribution of the pile cap. In fact, a structure (footing, beam, raft), connecting the pile heads and resting on the soil, produces important changes in the behaviour of the whole foundation.

Centrifuge tests by Horikoshi (1995) have shown that the load–settlement response of a single pile in clay is significantly affected by the presence of a small circular cap with a diameter only three times the diameter of the pile. Because of the contact pressure beneath the cap, the ultimate side resistance of the pile shaft is increased and the load–settlement relationship of the capped pile (Figure 10.1b) is different from that of the uncapped pile (Figure 10.1a) with no distinct collapse observed until the test was stopped at a settlement slightly lower than the diameter of the cap.

Fioravante *et al.* (2008) found essentially the same behaviour by centrifuge tests in loose saturated sand (Figure 10.2) comparing an isolated pile (not capped) with a piled raft (capped pile). Conte *et al.* (2003), again by centrifuge tests on square pile groups

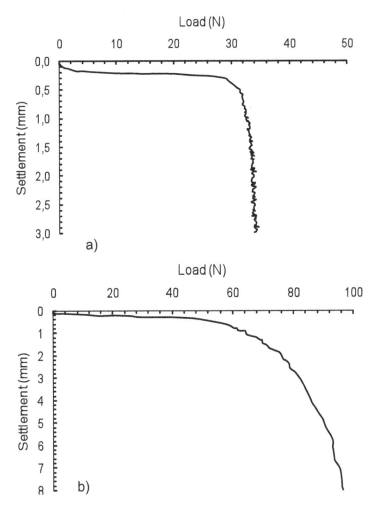

Figure 10.1 Centrifuge tests (a) uncapped pile, (b) capped pile (data from Horikoshi 1999).

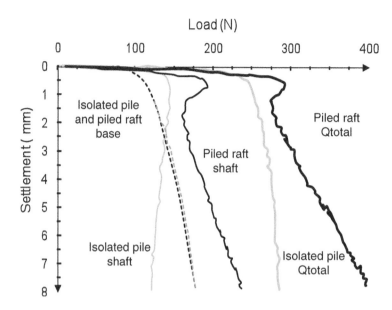

Figure 10.2 Centrifuge tests (data from Fioravante *et al.* 2008).

and piled rafts in clay, did not observe a distinct collapse for piled rafts. Numerical analyses (Reul 2004; de Sanctis and Mandolini 2006) support the above findings.

It can be concluded that even for small groups of piles the contact between the pile cap and the soil tends to produce a ductile behaviour and a punching type failure; of course, the effect is still more evident for large piled rafts. With such a behaviour, the focus of the design tends to move from bearing capacity to settlement.

According to codes and regulations, however, ultimate and serviceability limit states have always to be checked in design. With an outdated, but still widespread, language the design has to check that there is a sufficient margin against a bearing capacity failure; that the absolute and differential settlement are compatible with static and serviceability requirements of the structure; and that the foundation elements are proportioned to resist bending moments and shears. The methods of analysis are essentially those already presented in the previous chapters; improvements or integrations will be reported when needed.

One or the other of the above design requirements (limitation of absolute or differential settlement, sufficient margin against a bearing capacity failure, structural design) may be the critical one in various situations, depending on the characters of the structure and subsoil. The variety of possible situations is such that generalization is not yet always possible; accordingly, case histories or parametric studies will be presented, to grasp the influence of the different factors and trying to extract some general, albeit partial conclusions.

The presentation is essentially limited to vertical loads; only some preliminary comments will be provided for the case of horizontal loads.

10.2 Vertical bearing capacity

Safety against a bearing capacity failure is rarely a significant design problem for a raft foundation, even unpiled, in cohesionless soils. To support this statement, let us consider a raft resting at the surface of a cohesionless subsoil with a water table at the ground surface. In these rather unrealistic and extremely unfavourable conditions, if the width B of the raft is much smaller than its length L (plane strain conditions), the bearing capacity may be expressed as:

$$q_u = N_\gamma \gamma' B / 2$$

Assuming $B = 15\,\text{m}$, $\varphi' = 30°$ and hence $N_\gamma = 16.25$ (Davis and Booker 1971), $\gamma' = 10\,\text{kN/m}^3$, one obtains:

$$q_u = 16.25 \times 10 \times 7.5 = 1.22 \text{ MN/m}^2$$

With a global safety factor of the order of three, the service load could be over $400\,\text{kN/m}^2$, that is the load exerted by a residential or office building with around 40 floors. If the friction angle is even slightly higher than 30° and if the foundation plane and the ground water table are not at the soil surface but at a depth of even few metres, these value are easily doubled.

It is then evident that in cohesionless soils there are no bearing capacity problems for a raft foundation; even in the case of exceptionally heavy loads, it is easy to increase the bearing capacity by increasing the width and/or the depth of the foundation raft.

In the case of fine-grained soils in undrained conditions, on the contrary, the problem may exist if the soil is of low to medium consistency. In undrained conditions the bearing capacity, in terms of total stress, is given by:

$$q_u = N_c c_u$$

With $L >> B$, $N_c = 5.14$ and $c_u = 50\,\text{kN/m}^2$ (medium to stiff clay) we get $q_u = 257\,\text{kN/m}^2$ and hence $q_{all} = 86\,\text{kN/m}^2$ (a residential or office building with only around eight floors). It is again evident that, except for stiff or hard clay, the safety against bearing capacity can be a significant design problem.

In the case of stiff clay, the possible adoption of a piled raft is generally not intended as a means to improve the safety against a bearing capacity failure, since the raft alone will generally provide an adequate safety level, but rather to control the total or differential settlement and the stress in the raft. In the case of medium to soft clay, on the contrary, the addition of piles to the raft may be intended also to improve the safety against a bearing capacity failure; accordingly, the evaluation of the bearing capacity of a piled raft in clayey soils is still a relevant engineering problem.

The empirical expressions presented in §4.7, such as that suggested by Converse and Labarre or the Feld's rule and the block failure criterion suggested by Terzaghi and Peck, are conceived for pile groups and do not take into account the contribution of the raft.

Poulos (2000) states that the ultimate geotechnical capacity of a piled raft foundation may be estimated as the smaller of the following two values:

1   the sum of the ultimate capacity of the raft plus all the piles in the system (Liu *et al.* 1985);
2   the ultimate capacity of a block containing the piles and raft, plus that of the portion of the raft outside the periphery of the pile group.

Conventional design approaches, such as those reported in Chapter 4, can be used to estimate the various capacities in 1 and 2 above.

The installation of the piles may affect the soil properties and consequently modify the performance of the raft in comparison with that of the unpiled raft. Moreover, the behaviour of the piles belonging to a piled raft is affected not only by the interaction among piles but also by the surcharge exerted by the raft. As a consequence Liu *et al.* (1994) and Borel (2001) suggested expressing the bearing capacity of the piled raft Q_{PR} as follows:

$$Q_{PR} = \alpha_R \cdot Q_R + \alpha_P \cdot Q_P$$

where Q_R is the bearing capacity of the unpiled raft and Q_P that of the pile group, evaluated with an efficiency equal to unity, i.e. as the sum of the bearing capacity of the single piles. From a review of the available experimental evidence and a set of numerical analyses, de Sanctis and Mandolini (2006) conclude that α_p is always equal to unity, while α_R depends on the spacing and position of the piles and on the conventional displacement at failure considered. Defining the factor of safety of the unpiled raft $FS_R = \dfrac{Q_R}{Q}$, that of the pile group $FS_P = \dfrac{Q_P}{Q}$, where Q is the total load acting on the foundation, de Sanctis and Mandolini (2006) find that the factor of safety of the piled raft FS_{PR} is slightly lower than the sum of the two safety factors of the unpiled raft and the pile group. Specifically, they find:

$$0.82\left(FS_R + FS_P\right) \leq FS_{PR} \leq \left(FS_R + FS_P\right)$$

It is suggested that such a result be used in design.

10.3 Settlement

10.3.1 Introduction

From the viewpoint of the settlement control, it appears convenient to subdivide piled foundations in two broad groups: "small" and "large" piled rafts. To our purposes, they are defined as follows:

*   "small" piled rafts are those in which the bearing capacity of the unpiled raft would not be sufficient to carry the total load with a suitable factor of safety, and thus the primary reason to add piles is to achieve a sufficient factor of safety. This generally means that B amounts to a few metres, typically 5 to 15 m, and is small in comparison to the length L of the piles ($B/L < 1$). In this range, the raft may be made, and usually is, rather stiff and hence the differential settlement does not represent a major problem. A possible requisite for an optimum design is the limitation of the absolute settlement;

- "large" piled rafts are those in which the bearing capacity of the unpiled raft is sufficient to carry the total load with a suitable margin, so that the addition of piles is essentially intended to reduce settlement. In general, the width of the raft B is relatively large in comparison with the length of the piles ($B/L > 1$). For this reason, at least in relatively homogeneous soils, it may be impossible to significantly reduce the average settlement, as clearly explained by the sketch in Figure 10.3. Moreover, the raft cannot be but rather flexible; accordingly, a possible requisite for an optimum design is the limitation of differential settlement and bending moments and shear in the raft.

The limits given above between large and small rafts are tentative and largely conventional. Randolph (1994) claims that, to evaluate the settlement of a piled foundation, the method of the equivalent raft (presented in Chapter 5) has to be preferred when the modified aspect ratio $R = \sqrt{\dfrac{ns}{L}} \geq 4$, while the method of the equivalent pier (also presented in Chapter 5) is more suited if $R \leq 2$. It is interesting to note that, with the usual values of n and L, these limits are close to those suggested above between large and small rafts.

10.3.2 Small rafts

To explore the influence of some factors on the settlement of a small raft, let us consider the scheme reported in Figure 10.4: a square raft with $B = 15$ m acted upon by a uniform load q and resting on piles with a ratio s/d in the range from three to ten. The subsoil is modelled as an elastic half space; the ratio E_p/E_s has been taken equal to 1000. The parameters considered are: three values of the number of piles n (9, 25 or 49); three values of the ratio L/d (30, 60 and 90); three values of the ratio B/L (1, 0.5, 0.33); two values of the pile diameter d (0.5 m and 1 m). The raft is assumed to be rather stiff ($K_{rs} = \dfrac{4}{3}\dfrac{E_r\left(1-v^2\right)}{E\left(1-v_r^2\right)}\left(\dfrac{t}{B}\right)^3 = 1$), where E_r and v_r are the Young modulus and the Poisson's ratio and t the thickness of the raft.

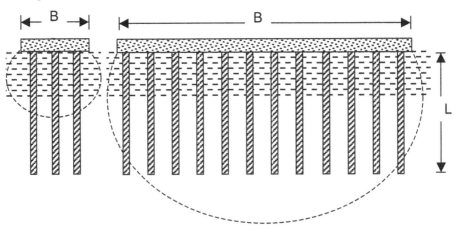

Figure 10.3 Small and large piled rafts.

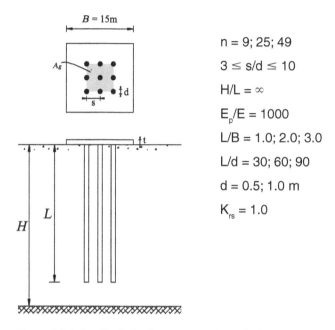

B = 15m

n = 9; 25; 49

3 ≤ s/d ≤ 10

H/L = ∞

E_p/E = 1000

L/B = 1.0; 2.0; 3.0

L/d = 30; 60; 90

d = 0.5; 1.0 m

K_{rs} = 1.0

Figure 10.4 Small piled raft – parametric analysis.

Let us call $A_g = [(\sqrt{n}-1)s]^2$ the area occupied by the piles, computed referring to the axis of the piles. $A_g \leq A = B^2$, where A is the total area of the raft. The ratio A_g/A characterizes the distribution of the piles below the raft. With a uniform distribution of the piles below the whole raft, A_g/A tends to unity (in practice, for small rafts, to values of 0.8 to 0.9). Considering a square group of piles centred below the raft, the values of A_g/A will be determined by the values of d, s and n.

The results obtained by a linearly elastic analysis are reported in Figures 10.5 and 10.6. The load Q_r directly transmitted by the raft to the soil, expressed as a

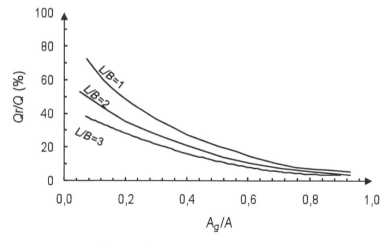

Figure 10.5 Load sharing for a small piled raft as a function of A_g/A.

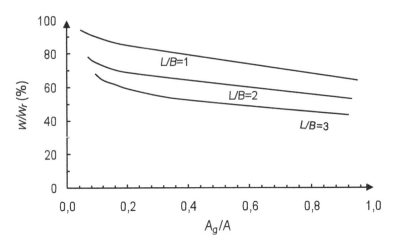

Figure 10.6 Settlement of a small piled raft as a function of A_g/A.

percentage of the applied load $Q=qB^2$, is plotted in Figure 10.5 as a function of A_g/A. Within the range of parameters considered, the load sharing between the piles and the raft is not significantly affected by L/d, s/d and n, but depends essentially on L/B and A_g/A.

The average settlement w of the piled raft, expressed as a percentage of the average settlement of the unpiled raft w_r, is reported in Figure 10.6 against A_g/A. Once again, the results show that the only parameters significantly affecting the settlement are B/L and A_g/A.

It is evident that, to reduce the average settlement, irrespective of the number and diameter of the piles it is convenient to have long piles uniformly spread below the foundation. This statement holds true even removing the simple assumptions of linearity and homogeneity with reference to the soil, taking into account that, for a "small" pile raft, the piles must be necessarily kept far from failure, being the main contributors to the overall safety factor of the foundation.

The case history of the pier of a cable-stayed bridge over the River Garigliano, already considered in §6.6, will be reported in some detail as an example.

The subsoil conditions at the site, reported in Figure 10.7 (Mandolini and Viggiani 1992), are characterized by a deep, rather compressible silty clay deposit. The foundation of the pier, resting on driven tubular steel piles 50 m long, is represented in Figure 10.8; according to the previous definition, it is a typical "small" piled raft.

Load tests to failure on instrumented piles and proof load tests on production piles were carried out (Bustamante *et al.* 1994; Russo 1996). The foundation was monitored during construction and afterwards, measuring settlement, load sharing between piles and raft, and load distribution among the piles.

The construction of the bridge started in October 1991 and the latest set of data has been recorded in October 2004, 13 years later. The settlement is measured by means of precision levelling; 35 out of the 144 piles were equipped with load cells at the top to measure the load transmitted by the cap to the pile; furthermore, eight pressure cells were installed at the interface between the cap and the soil. Further

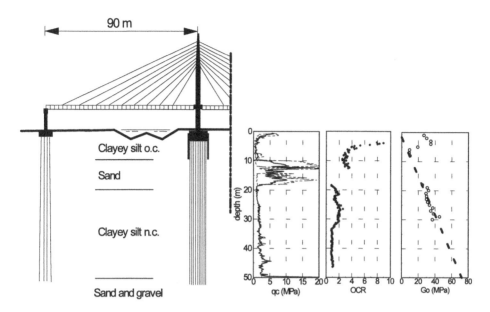

Figure 10.7 Subsoil profile at pier Number 7 Garigliano bridge.

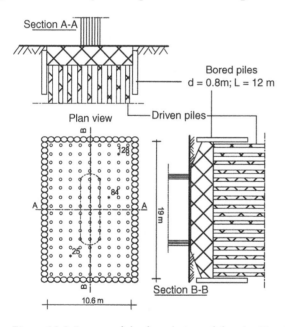

Figure 10.8 Layout of the foundation of the pier Number 7.

details on the instruments and the installation technique are reported by Mandolini *et al.* (1992) and Russo and Viggiani (1995).

In Figure 10.9 the load history and the measured average settlement are reported; differential settlement was negligible due to the very stiff pile cap. The net load is the total applied load minus the buoyancy, as deduced by piezometer readings.

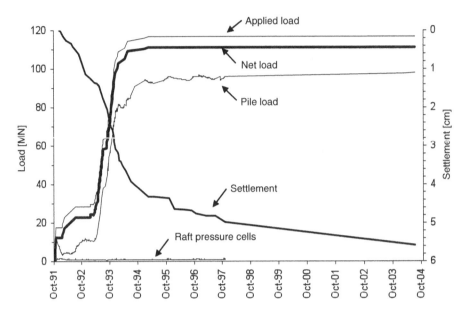

Figure 10.9 Total applied load and measured settlement for the foundation of the pier of the Garigliano bridge.

The total load acting on the foundation at the end of construction was $Q = 113\,\text{MN}$, approximately equal to the bearing capacity of the unpiled raft ($Q_R = 112\,\text{MN}$). In a conventional capacity based design, 144 piles were added in order to increase the bearing capacity. With an ultimate capacity of the single pile $Q_S = 3\,\text{MN}$, as deduced by load tests to failure, and a group efficiency equal to 0.7, the Italian regulations at the time of design (no contribution of the raft, $FS \geq 2.5$) have been satisfied (Viggiani 2001). Such a design resulted in a measured settlement of 52 mm, while the actual fraction of the total load transmitted to the piles was about 87% Q.

Russo (1996) back-analysed this case history by the code NAPRA, obtaining satisfactory agreement both for settlement (Figure 10.10) and for load distribution among the piles (Figure 10.11). The same numerical model was then adopted to predict the behaviour of the foundation with a decreasing number of piles uniformly spread below the raft. The results obtained are reported in Figure 10.12 in terms of settlement and load sharing.

Leaving apart the extreme solutions with a very small number of piles, for which the assumption of elastic raft–soil interaction adopted in NAPRA is questionable, Figure 10.12 shows that a significant reduction of pile number (say from 144 to 72) is possible without noticeable increase of the settlement. Within the same limits the prediction with the code Gruppalo neglecting the raft contributions is in substantial agreement with that of the code NAPRA.

From the point of view of the safety against failure, applying the suggestion of §10.2 one finds:

$$FS \geq 0.82(FS_P + FS_R) = 0.82\left(\frac{3 \times 72}{113} + \frac{112}{113}\right) = 2.38$$

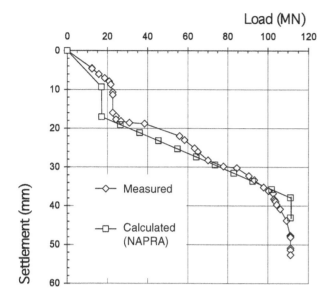

Figure 10.10 Pier Number 7 of the Garigliano bridge; comparison between predicted and observed settlement.

(a)

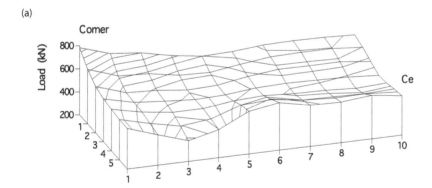

(b)

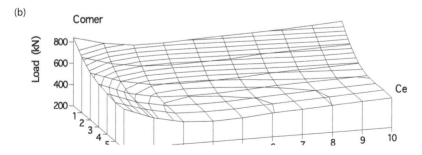

Figure 10.11 Pier Number 7 of the Garigliano bridge; comparison between predicted (b) and observed (a) load sharing.

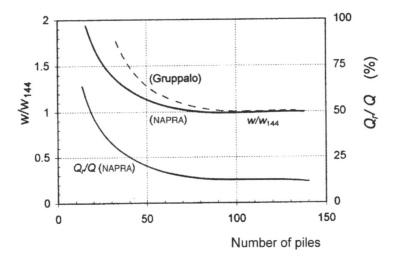

Figure 10.12 Pier Number 7 of the Garigliano bridge; settlement and load sharing as a function of the number of piles ($A_g/A \approx 1$).

It seems thus that there is a potential for substantial savings without significant reduction of performance.

10.3.3 Large rafts

A parametric study similar to the one reported in the previous section has been carried out for the large raft represented in Figure 10.13, for the range of parameters also reported in the figure. In this case the load–settlement relationship of the piles is modelled with a hyperbola: the initial stiffness is defined by using elastic solutions while the asymptote is defined via the α-method defined in Chapter 4 for piles in clayey soils.

The load sharing between raft and piles is reported in Figure 10.14 as a function of the different parameters. Unlike the case of the small raft (see Figure 10.5) the load sharing is markedly affected by all the parameters considered, with the only exception of the relative stiffness K_{rs} of the raft. Such a marked variation of the load sharing, however, affects to a much lesser degree the settlement w, reported in Figure 10.15 as a ratio to the settlement of the unpiled raft w_r. Indeed the variation of the average settlement does not exceed 25%, while the load share by the piles varies almost between 0% and 100% of the applied load. These results, and many similar ones not presented for brevity, lead to the conclusion already recalled on an intuitive basis (Figure 10.3): the addition of piles has not had a marked effect on the average settlement of a large raft founded on a deep compressible layer, and this essentially for the impossibility of providing sufficiently long piles.

The picture is completely different if the attention is shifted from the average to the differential settlement. In the following analyses the differential settlement δ between the centre and the corner of the raft will be considered; for the case of uniform load here examined, it is also the maximum differential settlement.

For the unpiled raft and for the two considered values of the flexural stiffness K_{rs}, the values of the average settlement w_r and the maximum differential settlement

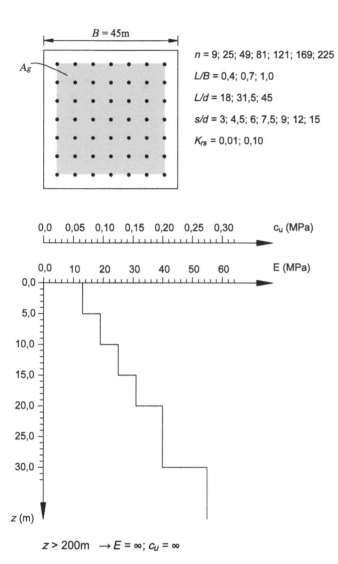

$n = 9; 25; 49; 81; 121; 169; 225$

$L/B = 0,4; 0,7; 1,0$

$L/d = 18; 31,5; 45$

$s/d = 3; 4,5; 6; 7,5; 9; 12; 15$

$K_{rs} = 0,01; 0,10$

$z > 200m \rightarrow E = \infty; c_u = \infty$

Figure 10.13 Large piled raft: case considered in the parametric study.

between centre and corner δ_r are reported in Table 10.1. The absolute and differential settlement are rather high, such that they are probably not compatible with the integrity of the superstructure.

Table 10.1 Average and differential settlement for the unpiled raft of Figure 10.13 under a load $Q = 800\,MN$

K_{rs}	w_r (mm)	δ_r (mm)
0.01	270	183
0.10	262	94

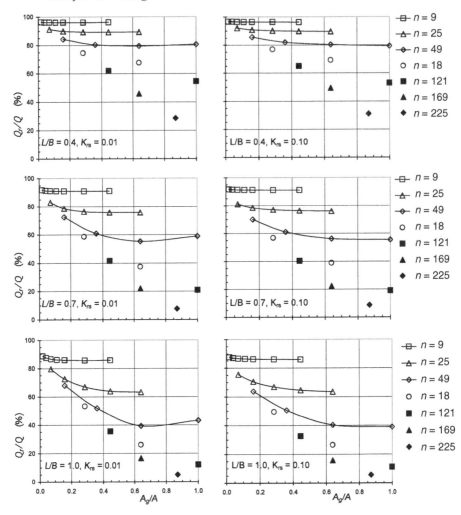

Figure 10.14 Load sharing for a large piled raft as a function of A_g/A.

Changing from a small to a large value of the relative stiffness K_{rs} the average settlement is practically unaffected and the differential settlement, though halved, remains large. It can be observed that to increase K_{rs} from 0.01 to 0.10 the thickness of the raft has to be increased from 1.1 to 2.3 m adding 2430 m^3 of concrete and increasing the weight by 60 MN. The variation is thus significant, probably near the upper limit of practical feasibility. This indicates that, for large rafts, the flexural stiffness of the raft does not exert a great influence.

The consequences on the differential settlement of the addition of piles will be explored by a broad parametric study, expressing the differential settlement between the corner and the centre δ of the piled raft as a fraction of the corresponding value δ_r of the unpiled raft.

The values of the ratio δ/δ_r are reported in Figure 10.16 for the different foundation layouts considered, as a function of the total pile quantity nL. For each of the

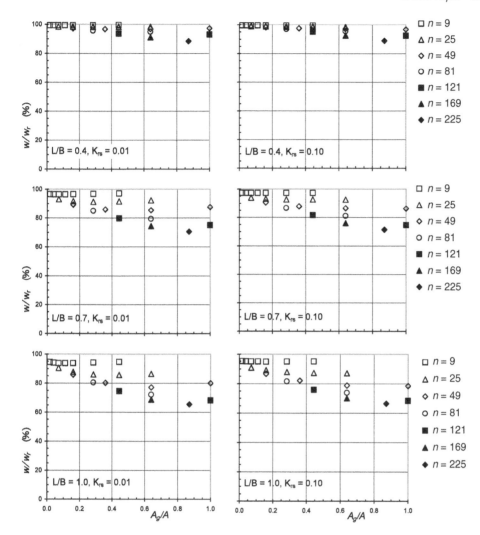

Figure 10.15 Settlement of a large piled raft as a function of A_g/A.

three values of the pile length L considered, the differential settlement is minimum when $s = 3d$; for the sake of clarity, Figure 10.17 refers to this value alone.

A scrutiny of the results leads to the following conclusions:

- the addition of piles to the raft is very effective in reducing the differential settlement. The value of δ may reduce to zero or even change sign, indicating that the corners of the raft are settling more than the centre;
- the longer the piles, the more effective they are in reducing the differential settlement. For the same total quantity of piles, a small number of long piles is the most convenient option;
- for each value of the pile length L, an optimum value of the pile quantity nL exists, corresponding to the maximum reduction of the differential settlement.

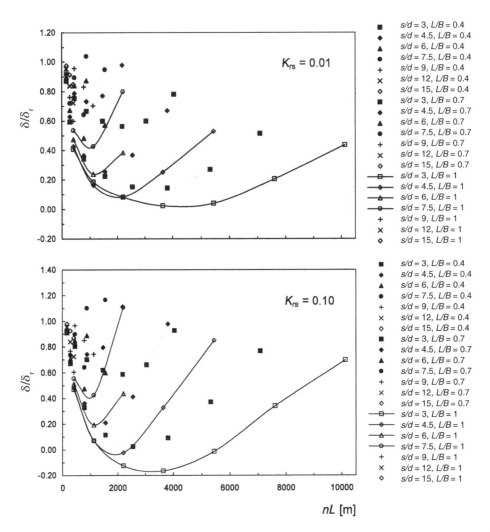

Figure 10.16 Differential settlement for a large piled raft as a function of the total length of the piles.

This occurs for values of A_g/A in the range 0.3 to 0.4 (group of few piles in the central zone of the foundation). Of course, this conclusion applies only for the case of uniformly distributed load; different pile layouts have to be explored to optimize the design for different load distributions.

As an example of large raft under approximately uniform load the building at Stonebridge Park (Cooke *et al.* 1981) will be presented; it is founded on London Clay, with $13 \times 27 = 351$ bored piles, 0.45 m in diameter and 13 m long, connected by a raft in contact with the ground (Figure 10.18).

Adopting a mean value of the undrained strength of the clay $c_u = 150$ kPa, the total bearing capacity of the unpiled raft may be evaluated to be around 700 MN, while the total permanent load is about 156 MN. The unpiled raft could hence carry the

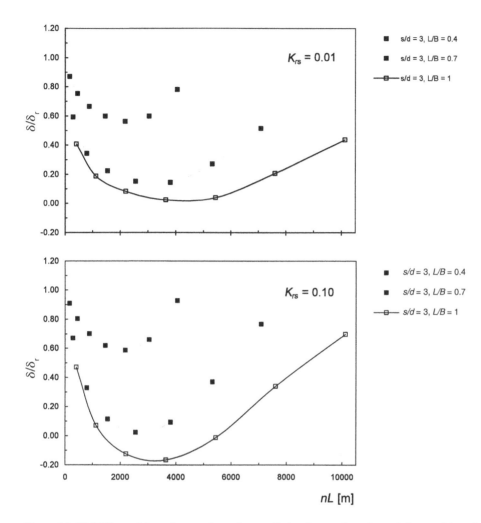

Figure 10.17 Differential settlement for a large piled raft as a function of the total length of the piles (only $s/d = 3$).

actual load with a suitable safety factor, and the piles have been added probably for absolute and differential settlement reduction. The bearing capacity of a single pile, as assessed by load tests, is around 1.6 MN; accordingly, the whole load can be carried entirely by the piles, with a safety factor of $(0.8 \times 351 \times 1.6)/156 = 2.9$, having assumed a group effect on bearing capacity equal to 0.8. This is a typical example of traditional, capacity based design.

The observed settlement are compared in Figure 10.19 to those predicted by NAPRA in both undrained and drained conditions; the prediction is based on the results of two load tests on piles. It may be seen that the predicted undrained settlement (11 mm) matches well the end-of-construction measured one, while the predicted drained settlement (18 mm) fits the observed final settlement. The calculated maximum differential settlement between the corner and the centre of the raft is 13 mm.

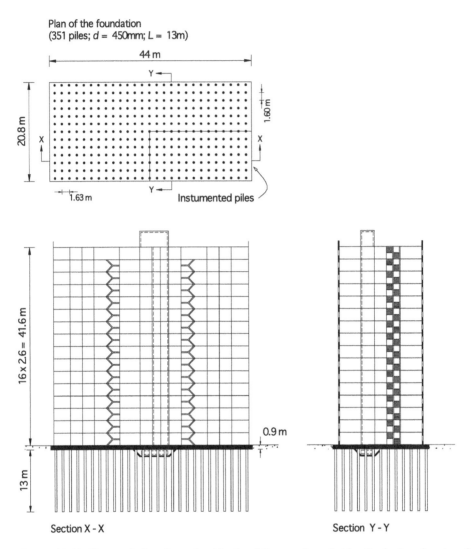

Plan of the foundation
(351 piles; d = 450mm; L = 13m)

Figure 10.18 Plan and elevation of a block of flats at Stonebridge Park, London (modified
after Cooke *et al.* 1981).

On the basis of measurements of the load at the head of some piles and of the contact pressure between the raft and the soil, it has been found that in the early stage of construction the raft carried above 40% of the total applied load; this percentage decreased below 25% at the end of the observation period.

The calculations carried out by NAPRA (Russo 1996; Mandolini *et al.* 1997; Russo 1998b; Viggiani 1998) in undrained conditions gave a load on the raft equal to about 20% of total; under drained conditions the load on the raft was almost negligible.

The overall agreement between prediction and observation is rather satisfactory.

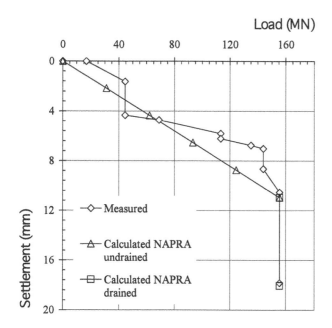

Figure 10.19 Predicted and observed settlement of the building at Stonebridge Park, London.

In the case of Stonebridge Park, since the unpiled raft alone has a sufficient bearing capacity, the piles could have been better used to control the absolute or differential settlement, with a substantial saving in their number. To substantiate this statement, the foundation has been redesigned by NAPRA, with different numbers and layouts of piles. The results obtained are reported in Figures 10.20 and 10.21.

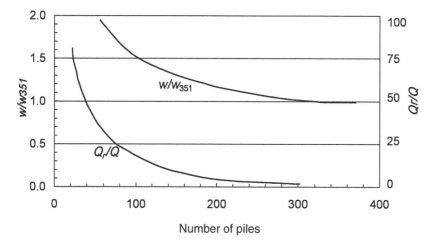

Figure 10.20 Stonebridge Park: increase of the average settlement with decreasing the number of piles. $A_g/A = 88\%$.

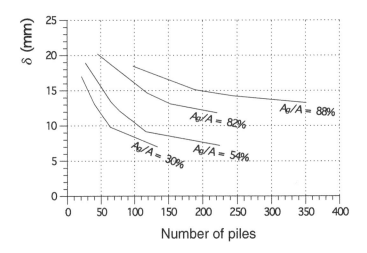

Figure 10.21 Stonebridge Park: effect of the number and layout of the piles on the maximum differential settlement.

The effect on the average settlement of the reduction of the number of piles, keeping the ratio $A_g/A = 88\%$ (i.e. keeping the piles uniformly spread over the whole foundation area) is reported in Figure 10.20. The average settlement increases with the decreasing number of piles, but at a rather slow rate. Halving the number of piles from 351 to 175 increases the average settlement by only 12.5%; with 117 piles, i.e. one-third of the actual number, the settlement increases by 50%. In the mean time (Figure 10.21) the maximum differential settlement increases respectively by 17% and 32%.

The differential settlement, however, may be decreased dramatically by concentrating the piles in the central area of the raft, i.e. by decreasing the value of the ratio A_g/A. Figure 10.21 shows that, with $A_g/A = 30\%$, only 40 piles are sufficient to get the same differential settlement of the case $A_g/A = 88\%$, n = 351. For large rafts, the potential for a more rational design with substantial economy is evident.

An example of non-uniform load distribution is reported by de Sanctis *et al.* (2002), with an exercise of optimization of the foundation of the two towers in the eastern area of Naples already discussed in §5.4.

As reported in §5.4, the code NAPRA was used to interpret the observed behaviour of the towers in terms of settlements. The plan and the cross-section of the (one-quarter) foundation model used in the analysis are reported in Figure 10.22, together with the location of the distributed and concentrated loads.

The measured and computed settlement for the actual piled raft and for the unpiled raft is plotted in Figure 10.23. The provision of 637 piles uniformly spread below the raft, resulting from a capacity-based design, reduces the average settlement by around 30% but is much less effective in reducing the differential settlement. Such a result can be considered typical of large piled rafts where the ratio between the raft width and the pile length $B/L > 1$; from a practical point of view, this occurs when the foundation width B is of the order of some tens of metres.

The agreement between computed and observed settlement is quite satisfactory, and thus the same computational model has been used to redesign the foundation

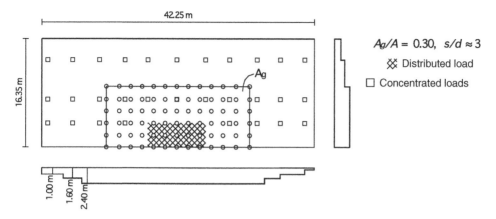

Figure 10.22 Foundation model adopted in the analysis for the Holiday Inn building.

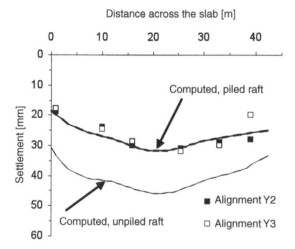

Figure 10.23 Measured and computed settlement for the actual piled and unpiled raft.

with different criteria (small number of piles concentrated under the stiffening cores). de Sanctis *et al.* (2002) demonstrated that the total pile length $nL \approx 12\,700$ m (and hence, very nearly, the cost of the foundation) may be halved with a 25% reduction of the differential settlement and only a 10% increase of the maximum settlement.

10.4 Horizontal load

The insight of the mechanisms controlling interaction between the three components of a piled foundation (cap, piles and soil) subjected to vertical loads, gained by a broad amount of theoretical and experimental investigations, allowed the evolution of new design criteria, already applied in quite a number of important projects. For horizontal loading the situation is substantially different: the experimental evidence, as reviewed for instance in Chapter 9 of this book, is much more limited and scanty. Furthermore, many of the available investigations are out of date and only very recently a new interest in this topic may be found in the literature. For such reasons, only a few general comments will be reported here, based on the uncertain support of the few and somewhat contradictory experimental and theoretical studies available.

Horizontal load on a piled raft foundation are resisted by:

i   the piles;
ii   the passive resistance of the soil on the front of the embedded structure;
iii   the frictional resistance along the embedded sides;
iv   the frictional resistance along the base of the raft.

In the conventional design approach only the first contribution is generally taken into account by the theories and methods described in Chapters 8 and 9 with reference to pile groups. Much like what occurs for vertical loads, such an approach may be overly conservative since it neglects the additional resistance provided by the raft–soil system as sum of the contributions (ii), (iii) and (iv) above.

On the other hand, in general, the above resistances develop at different displacement so that progressive failure starting from the stiffer component could occur; the situation could be particularly dangerous should the stiffer component be the piles.

A number of investigations have confirmed that, as was to be expected, the stiffness and bearing capacity of a piled raft under horizontal load are greater than those of a free-standing pile group.

Mokwa (1999) reports lateral loading experiments on a small square raft with four driven steel H piles with length in the range from 3 to 6 m, at a spacing $4d = 1$ m, being $d = 250$ mm the diameter of a circular section equivalent the H pile. Figure 10.24 reports some of the results, clearly showing the positive influence of the cap embedment.

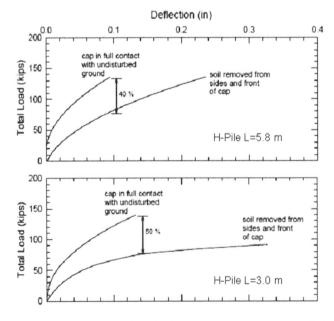

Figure 10.24 Measured load–displacement response for piled raft under horizontal load (after Mokwa 1999).

Similar results have been obtained by centrifuge test (Horikoshi *et al.* 2002), small scale laboratory tests (Katzenbach and Turek 2005), full-scale field tests (Beatty 1970; Kim and Sing 1979; Rollins *et al.* 1998; Zafir and Vanderpool 1998) and by numerical analyses (Cunha and Zhang 2006).

Katzenbach and Turek (2005) performed model tests on a piled raft resting at the soil surface (zero embedment) and on a free-standing pile group; it is thus possible to isolate the contribution of the soil friction on the raft (item (iv) above). In Figure 10.25 the ratio between the load taken by the cap only and the total applied load is plotted against the horizontal displacement. Figure 10.25 shows that, in the case considered, the contribution may be as high as 40% at failure and increases with increasing the vertical load acting on the raft. In the early stages of the lateral loading, and hence for small horizontal displacements, a major part of the horizontal load is carried by the friction on the raft; this portion decreases when the displacement increases.

While the beneficial effect of the raft–soil contact in terms of stiffness and bearing capacity appears generally acknowledged and fully confirmed by the available evidence, the effect of the same contact on the bending moments in the piles is less understood and the available evidence is scarce and somewhat contradictory.

Kim and Sing (1979) performed lateral load experiments on two equal pile groups, one free standing and another one with the raft in contact with the soil. In the early stages of the test the maximum bending moments observed in the piles were very similar in the two experiments; this finding has been explained with a negligible mobilization of the friction between the raft and the soil. At higher load level, however, the observed moments in the piles of the raft in contact were less than half those in the free standing group, due to the contribution of the friction between the raft and the soil resulting in a decrease of the loads transmitted to the pile head.

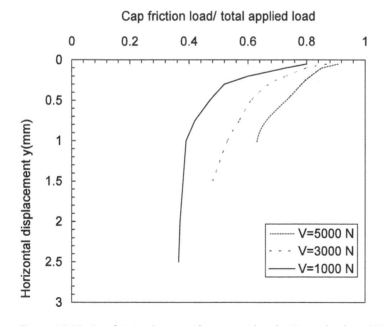

Figure 10.25 Cap friction by centrifuge tests (data by Katzenback and Turek 2005).

On the other hand Horikoshi *et al.* (2002), in a centrifuge test, observed that the raft–soil friction is fully mobilized at a displacement much lower than that needed to mobilize the lateral bearing capacity of the piles.

Some attempts have been made to grasp the phenomenon under generalized loading by numerical analyses (e.g. Cunha and Zhang 2006), but further investigations on the effect of a raft in contact with the soil are still badly needed.

10.5 Raft foundations with disconnected piles

While it is common practice for pile heads to be structurally connected with the raft, recently Wong *et al.* (2000) suggested an alternative design concept for settlement reducing piles, in which the piles are not structurally connected to the raft but a gap is left in between. They argue that, for an efficient design of rafts with settlement reducing piles, under working load conditions the axial load in the piles should be as high as 80% or more of their ultimate geotechnical capacity, while the performance of the foundation is still satisfactory because of the contribution of the raft. In these cases the design of the piles could be governed by their structural, rather than geotechnical, capacity.

Legend
1. Caisson
2. Flood gate
3. Ballast (water)
4. Ballast (concrete)
5. Plant and access tunnels
6. Sheet pile
7. Settlement reducing piles

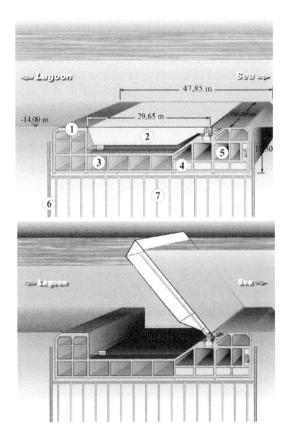

Figure 10.26 MOSE barriers in Venice against high tide founded on disconnected driven piles.

In addition, the relatively few piles beneath the raft may not provide adequate resistance against lateral loads, and in any case very high bending moments occur at the connection between piles and raft. Finally, when the piles are connected to the raft, they have to be considered as a structural element and designed according to the factor of safety prescribed for such elements by codes and regulations.

If the piles are not connected to the raft, the bending moments and shear induced by lateral loads substantially decrease, while the lateral load acting on the structure is resisted by friction at the interface between the raft and the soil, and possibly by passive pressure on the basement walls. Also the maximum axial force in each pile decreases slightly, due to the mechanism of load transfer by negative friction on the upper part of the pile shaft. Randolph (1983) reports that even the localized high bending moment in the raft near the piles are reduced if the piles are disconnected from the raft. Furthermore, once the piles are no longer connected to the raft, they may be viewed as soil reinforcing elements, and a much lower factor of safety against structural failure than that suggested for structural piles can be used without violation of most codes.

Some applications of a disconnected settlement reducing pile are beginning to appear in the literature. Thornburn *et al.* (1983) report the case of the foundations of a storage tank; in this case, however, over 90% of the total tank load has been transferred to the piles; hence the situation is somewhat different from that of settlement reducing piles.

Burghignoli *et al.* (2007) and Jamiolkowski *et al.* (2009) report the case of the MOSE mobile barriers system, currently under construction as the most important and challenging part of the interventions to safeguard Venice against high tide. The caissons of the mobile barriers, housing flap gates ensuring the closure of the lagoon inlets when a high tide of elevation +1.10 m or higher is forecasted (Figure 10.26), are founded on disconnected precast driven piles.

References

Abagnara, V. (2009) *Modelling and Analysis of Single Pile and Pile Groups under Horizontal Load*, PhD thesis, University of Napoli Federico II [in Italian].

Abagnara, V. and Russo, G. (2007) *Analisi del comportamento di pali di fondazione sollecitati trasversalmente all'asse*, IARG Luglio, Salerno, Italy.

Alizadeh, M. and Davisson, M.T. (1970) Lateral load tests on piles: Arkansas River Project, *Journal of the Soil Mechanics and Foundations Division, ASCE*, 96(50: 1583–1604.

Agerschou, H.A. (1962) Analysis of the Engineering News pile formula, *Journal of the Soil Mechanics and Foundations Division, ASCE*, 88(8): 13–34.

Amir, J.M. (1986) *Piling in Rock*, Balkema, Rotterdam.

Aoki, N. and Velloso, D.A. (1975) An approximate method to estimate the bearing capacity of piles, *Proceedings of the Fifth Pan-American Conference of Soil Mechanics and Foundation Engineering, Vol. 1*, Buenos Aires, pp. 367–376.

Aoki, N., Velloso, D.A. and Salamoni, J.A. (1978) Fundações para o silo vertical de 100000 t no porto de Paranaguà, *Sixth Brazilian Conference of Soil Mechanics and Foundation Engineering, Vol. 3*, Rio de Janeiro, pp. 125–132.

API (1993) *Recommended Practice for Planning, Designing and Constructing Fixed Offshore Platforms: Working Stress Design*, API RP 2A-WSD, twentieth edition, American Petroleum Institute, Washington, DC.

Aravin, V.L. and Numerov, S.L. (1964) *Theory of Fluid Flow in Undeformable Porous Media*, Israel Program for Scientific Translations, Jerusalem.

Banerjee, P.K. and Davies, T.G. (1978) The behaviour of axially and laterally loaded single piles embedded in non homogeneous soils, *Géotechnique*, 28(3): 309–326.

Banerjee, P.K. and Driscoll, R.M. (1978) *Program for the Analysis of Pile Groups of any Geometry Subjected to Horizontal and Vertical Loading and Moments: PGROUP 2.1*, HECB/B/7, Department of Transport, HECB, London.

Barton, Y.O. (1984) Response of pile groups to lateral loading in the centrifuge, *Proceedings of the Symposium on the Applications of Centrifuge Modelling to Geotechnical Design*, Balkema, Rotterdam.

Bear, J. (1972) *Dynamics of Fluids in Porous Media*, Elsevier, New York.

Beatty, C.I. (1970) Lateral test on pile groups, *Foundation Facts*, 6(1): 18–21.

Becker, D.E. (1997) Eighteenth Canadian Geotechnical Colloquium: limit states design for foundations, *Canadian Geotechnical Journal*, 33(6): 956–983.

Berezantsev, V.G. (1965) Design of deep foundations, *Proceedings of the Sixth International Conference of Soil Mechanics and Foundation Engineering, Vol. 2*, Montreal, pp. 234–237.

Berezantsev, V.G., Kristoforov, V. and Golubkov, V. (1961) Load bearing capacity and deformation of pile foundations, *Proceedings of the Fifth International Conference of Soil Mechanics and Foundation Engineering, Vol. 2*, Paris, pp. 11–15.

Bermingham, P. and Janes, M. (1989) An innovative approach to load testing of high capacity piles, *Proceedings of the International Conference on Piling and Deep Foundations*, London, pp. 409–413.

Bilotta, E., Caputo, V. and Viggiani, C. (1991) Analysis of soil structure interaction for piled rafts, *Proceedings of the Tenth European Conference of Soil Mechanics and Foundation Engineering, Vol. 1*, Firenze, pp. 315–318.

Biot, M.A. (1941) General theory of three dimensional consolidation, *Journal of Applied Physics*, 12: 155–164.

Bishop, R.F., Hill, R. and Mott, N.F. (1945) The theory of indentation and hardness tests, *Proceedings of the Physical Society*, 57(3): 147–159.

Bishop, A.W. and Bjerrum, L. (1961) The relevance of the triaxial test to the solution of stability problems, *Proceedings, ASCE Research Conference on Shear Strength of Cohesive Soils*, Boulder, CO, pp. 437–501.

Borel, S. (2001) *Comportement et dimensionnement des fondations mixtes*, These de Docteur de ENPC, Spécialité Géotechnique, Paris.

Boscardin, M.D. and Cording, E.J. (1989) Building response to excavation induced settlement, *Journal of Geotechnical Engineering, ASCE*, 115(1): 1–21.

Boussinesq, M.J. (1885) *Applications des potentiels à l'étude de l'équilibre et de mouvements des solides élastiques*, Gauthier Villars, Paris.

Bowles, J.E. (1988) *Foundation Analysis and Design*, fourth edition, McGraw Hill, New York.

Brand, E.W., Muktabhant, C. and Taechathummarak, A. (1972) Load tests on small foundation in soft clay, *Proceedings of the ASCE Conference Performance of Earth and Earth Supported Structures*, I(2): 903–928, Purdue University, West Lafayette, IN.

Brettmann, T. and Duncan, J.M. (1996) Computer application of CLM lateral load analysis to piles and drilled shafts, *Journal of Geotechnical Engineering, ASCE*, 122(6): 496–498.

Briaud, J.L.,Tucker, L.M. and Ng, E. (1989) Axially loaded five pile group and single pile in sand, *Proceedings of the Twelfth International Conference of Soil Mechanics and Foundation Engineering, Vol. 2*, Rio de Janeiro, pp. 1121–1124.

Brinch Hansen, J. (1951) *Simple Statical Computation of Permissible Pile Load*, CN-Post, Copenhagen.

Brinch Hansen, J. (1956) *Limit State and Safety Factors in Soil Mechanics*, Danish Geotechnical Institute, Copenhagen, Bulletin No. 1 [in Danish with an English summary].

Broms, B.B. (1964a) Lateral resistance of piles in cohesive soils, *Journal of Soil Mechanics and Foundation Engineering, ASCE*, 90(2): 27–63.

Broms, B.B. (1964b) Lateral resistance of piles in cohesionless soils, *Journal of Soil Mechanics and Foundation Engineering, ASCE*, 90(3): 123–156.

Brown, D.A. and Shie, C.F. (1990a) Three-dimensional finite element model of laterally loaded pile, *Computers and Geotechnics*, 10(1): 59–79.

Brown, D.A. and Shie, C.F. (1990b) Numerical experiments into group effects on the response of piles to lateral loading, *Computers and Geotechnics*, 10(3): 211–230.

Brown, D.A., Morrison, C. and Reese, L.C. (1988) Lateral load behavior of pile group in sand, *Journal of Geotechnical Engineering, ASCE*, 114(11): 1261–1276.

Brown, D.A., Reese, L.C. and O'Neill, M.W. (1987) Cyclic lateral loading of a large scale pile group, *Journal of Geotechnical Engineering, ASCE*, 113(11): 1326–1343.

Brown, M.J. and Hyde, A.F.L. (2008) Rate effects from pile shaft resistance measurements, *Canadian Geotechnical Journal*, 45(3): 425–431.

Burghignoli, A., Jamiolkowski, M. and Viggiani, C. (2007) Geotechnics for the preservation of historic cities and monuments: components of a multidisciplinary approach, *Proceedings of the Fourteenth European Conference on Soil Mechanics and Geotechnical Engineering*, Millpress, Amsterdam, pp. 3–38.

Burland, J.B. (1973) Shaft friction on piles in clay: a simple fundamental approach, *Ground Engineering*, 6(3): 30–42.

Burland, J.B. and Burbidge, M.C. (1985) Settlement of foundations in sand and gravel, Centenary Lecture, Glasgow and West Scotland Association, reproduced in *ICE Proceedings*, 78(6): 1325–1381.

Burland, J.B. and Wroth, C.P. (1974) Allowable and differential settlement of structures, including damage and soil structure interaction, *BGS Conference on Settlement of Structures*, Pentech Press, Cambridge, pp. 611–673.

Burland, J.B., Broms, B.B. and De Mello, V.F.B. (1977) Behaviour of foundations and slopes, *Proceedings of the Ninth International Conference on Soil Mechanics and Foundation Engineering*, I: 495–546, Tokyo.

Bustamante, M. and Doix, B. (1985) Une mèthode pour le calcul des tirants et des micropieux injectés, *Bulletin Liaison des Laboratoires Ponts et Chaussées*, 140: 75–95.

Bustamante, M., Gianeselli, L., Mandolini, A. and Viggiani, C. (1994) Loading tests on slender driven piles in clay, *Proceedings of the Thirteenth International Conference on Soil Mechanics and Foundation Engineering, Vol. 2*, New Delhi, p. 685.

Butler, F.G. (1974) Settlement in heavily overconsolidated clay, *BGS Conference on Settlement of Structures*, Pentech Press, Cambridge, pp. 531–578.

Butterfield, R. and Banerjee, B.K. (1971) The elastic analysis of compressible piles and pile groups, *Géotechnique*, 21(1): 43–60.

Callisto, L. (1994) Experimental observations on two laterally loaded bored piles in clay, *Proceedings of a Workshop on Pile Foundations: Experimental Investigations, Analysis and Design*, Napoli, pp. 349–359.

Caputo, V. and Viggiani, C. (1984) Pile foundations analysis: a simple approach to nonlinearity effects, *Rivista Italiana di Geotecnica*, 18(1): 32–51.

Caquot, A. (1934) *Equilibre des massiffs à frottement interne*, Gauthier Villars, Paris.

Caquot, Q.A. and Kerisel, J. (1956) *Traité de Mécanique des Sols*, Gauthier Villars, Paris.

Cedergreen, H.R. (1967) *Seepage, Drainage and Flow Nets*, John Wiley and Sons, New York.

Chang, M.F. and Wong, I.H. (1987) Shaft friction of drilled piers in weathered rock, *Proceedings of the Sixth International Conference on Rock Mechanics*, Montreal.

Chin, F.K. (1970) Estimation of the ultimate load of piles from tests not carried to Failure, *Proceedings of the Second South East Asian Conference on Soil Engineering*, Singapore, pp. 81–92.

Chow, F.C. (1997) *Investigation in the Behaviour of Displacement Piles for Offshore Foundations*, PhD thesis, Imperial College, London.

Christian, J.T. and Carrier III, W.D. (1978) Janbu, Bjerrum and Kjaernsli's chart reinterpreted, *Canadian Geotechnical Journal*, 15(1): 124–128.

Conte, G., Mandolini, A. and Randolph, M.F. (2003) Centrifuge modelling to investigate the performance of piled rafts, *Proceedings of the Fourth International Conference on Foundations on Bored and Auger Piles*, BAP IV, pp. 359–366.

Cooke, R.W. (1974) Settlement of friction pile foundations, *Proceedings of the Conference on Tall Buildings*, Kuala Lampur, p. 7.

Cooke, R.W. (1986) Piled raft foundations on stiff clays: a contribution to design philosophy, *Géotechnique*, 36(2): 169–203.

Cooke, R.W., Bryden Smith, D.W., Gooch, M.N. and Sillet, D.F. (1981) Some observations on the foundation loading and settlement of a multi-storey building on a piled raft foundation in London Clay, *ICE Proceedings*, 70(3): 433–460.

Coulomb, C.A. (1776) *Essai sur une application des règles de maximise t minimis à quelques problèmes de statique rélatifs à l'architecture*, Mémoires de Mathématique et de Physique presents à l'Academie des Sciences, Paris, pp. 343–382.

Cundall, P.A. and Strack, O.D.L. (1979) Discrete numerical model for granular assemblies, *Géotechnique*, 29(1): 47–65.

Cunha, R.P. and Zhang, H.H. (2006) Behavior of piled raft foundation systems under a combined set of loadings, *Proceedings of the Tenth International Conference on Piling and*

Deep Foundations, J. Lindenberg, M. Bottiau and A.F. van Tol (eds), Amsterdam, pp. 242–251.

d'Arcy, H. (1856) *Les fontaines publiques de la ville de Dijon*, Dunod, Paris.

Davies, T.G. and Budhu, M. (1986) Non-linear analysis of laterally loaded piles in heavily overconsolidated clays, *Géotechnique*, 36(4): 527–538.

Davis, E.H. and Booker, J.R. (1971) The bearing capacity of strip footing from the standpoint of plasticity theory, *Proceedings of the First Australia and New Zealand Conference on Geomechanics, Vol. 1*, Sydney, pp. 276–282.

Davis, E.H. and Taylor, H. (1961) The surface displacement of an elastic layer due to horizontal and vertical surface loading, *Proceedings of the Fifth International Conference on Soil Mechanics and Foundation Engineering, Vol. 1*, Paris, p. 621.

Davisson, M.T. (1970) Lateral load capacity of piles, *Highway Research Record*, 333: 104–112.

Davisson, M.T. and Prakash, S. (1963) A review of soil–pile behavior, *Highway Research Record*, 39: 25–48.

De Beer, E. (1945) Etude des fondations sur piloitis et des fondations directes, *Annales des Travaux Publiques de Belgique*, 46: 1–78.

De Cock, F. and Legrand, C. (eds) (1997) Design of axially loaded piles: European practice, *Proceedings of the ERTC3 Seminar*, Balkema, Rotterdam.

Decourt, L. (1995) Prediction of load–settlement relationships for foundations on the basis of SPT, *Ciclo de Conferencias Internationale "Leonardo Zeevaert"*, UNAM Mexico, pp. 85–104.

de Sanctis, L. and Mandolini, A. (2006) Bearing capacity of piled rafts on soft clay soils, *Journal of Geotechnical and Geoenvironmental Engineering*, ASCE, 132: 1600–1610.

de Sanctis, L., Mandolini, A., Russo, G. and Viggiani, C. (2002) *Some Remarks on the Optimum Design of Piled Rafts*, Deep Foundation Congress, GeoInstitute of the ASCE, Orlando, FL.

Douglas, D.J. and Davis, E.H. (1964) The movement of buried footings due to moment or horizontal load and the movement of anchor plates, *Géotechnique*, 14(2): 115–132.

Duncan, J.M., Evans, L.T. and Ooi, P.S.K. (1994) Lateral load analyses on single piles and drilled shafts, *Journal of Geotechnical Engineering*, ASCE, 120(6): 1018–1033.

Dunnicliff, J. and Young, N.P. (2006) *Ralph B. Peck, Educator and Engineer: The Essence of the Man*, BiTech Publishers Ltd, Vancouver.

El Sharnouby, B. and Novak, M. (1986) Flexibility coefficients and interaction factors for pile group analysis, *Canadian Geotechnical Journal*, 23(4): 441–450.

Ellison, R.D., D'Appolonia, E. and Thiers, G.R. (1971) Load–deformation mechanism for bored piles, *Journal of the Soil Mechanics and Foundation Engineering Division*, ASCE, 97(4): 661–678.

El Mossallamy, Y. and Franke, E. (1997) *Piled Rafts: Numerical Modeling to Simulate the Behaviour of Piled Raft Foundations*, published by the authors, Darmstadt.

Eslami, A. and Fellenius, B.H. (1997) Pile capacity by direct CPT and CPTu methods applied to 102 case histories, *Canadian Geotechnical Journal*, 34(6): 886–904.

Esposito, L., Nobile, L. and Viggiani, C. (1975) An approximate solution for displacements in a non-homogeneous elastic layer, *Proceedings of the Istanbul Conference on Soil Mechanics and Foundation Engineering*.

Evans, L.T. and Duncan, J.M. (1982) *Simplified Analysis of Laterally Loaded Piles*, Report UCB/GT/82–04, Department of Civil Engineering, University of California, Berkeley, CA.

Fischer, K. (1957) Zur Berechnung der Setzung einer starren mittig belasteten Kreisplatte auf geshichteter Unterlage, *Beton- and Stahlbetonbau*, H.10.

Fleming, W.G.K., Weltman, A.J., Randolph, M.F. and Elson, W.K. (1985) *Piling Engineering*, Surrey University Press/Halsted Press, Glasgow.

Frank, R.A. (1974) *Etude théorique du comportement des pieux sous charge verticale. Introduction de la dilatance*, Rapport de recherché No. 46, LCPC, Paris.

Franke, E. (1989) Co-report, Session 13: large diameter piles, *Proceedings of the Twelfth International Conference on Soil Mechanics and Foundation Engineering*, Rio de Janeiro.

Ghionna, V.N., Jamiolkowski, M., Pedroni, S. and Salgado, R. (1994) The tip displacement of drilled shafts in sand, *Proceedings "Settlement '94", Geotechnical Engineering Division, ASCE*, 2: 1039–1057.

Go, V. and Olsen, R.E. (1993) Axial load capacity of untapered piles in sand, *Proceedings of the Eleventh South East Asia Geotechnical Conference*, Singapore, pp. 517–521.

Golder, H.Q. and Osler, J.C. (1968) Settlement of a furnace foundation, Sorel, Quebec, *Canadian Geotechnical Journal*, 5(1): 46–56.

Grant, R., Christian, J.T. and Vanmarke, E.H. (1974) Differential settlement of buildings, *Journal of the Geotechnical Engineering Division, ASCE*, 100(9): 973–991.

Griffith, D.V., Clancy, P. and Randolph, M.F. (1991) Piled raft analysis by finite elements, *Proceedings of the Seventh International Conference on Computer Methods and Advances in Geomechanics*, Cairns, Queensland, 2: 1153–1157.

Hain, S. and Lee, I.K. (1978) The analysis of flexible raft-pile systems, *Géotechnique*, 28(1): 65–83.

Harr, M.E. (1966) *Foundations of Theoretical Soil Mechanics*, McGraw Hill, New York.

Hiley, A. (1925) A rational pile driving formula and its application in piling practice explained, *Engineering*, 119(3100): 657–658.

Horikoshi, K. (1995) *Optimum Design of Piled Raft Foundations*, PhD thesis, University of Western Australia.

Horikoshi, K., Matsumoto, T., Watanabe, T. and Fukuyama, H. (2002) Performance of piled raft foundations subjected to seismic loads, in Y. Honjo, O. Kusakabe, K. Matsui, M. Kouda and G. Pokharel (eds), *Foundation Design Codes and Soil Investigation in View of International Harmonization and Performance*, Swets & Zeitlinger, Lisse, pp. 381–389.

Horvath, R.G., Kenney, T.C. and Trow, W.P. (1980) Results of tests to determine shaft resistance of rock socketed drilled piers, *Proceedings of the International Conference on Structural Foundations on Rock*, 1: 349–361, Sydney.

Huang, A.B., Hsueh, C.K., O'Neill, M.W., Chern, S. and Chen, C. (2001) Effects of construction on laterally loaded pile groups, *Journal of Geotechnical and Geoenvironmental Engineering, ASCE*, 127(5): 385–397.

Hu, Y. and Randolph, M.F. (2002) Bearing capacity of caisson foundations on normally consolidated clay, *Soils and Foundations*, 42(5): 71–77.

Ilyas, T., Leung, C.F., Chow, Y.K. and Budi, S.S. (2004) Centrifuge model study of laterally loaded pile groups in clay, *Journal of Geotechnical and Geoenvironmental Engineering, ASCE*, 130(3): 274–283.

Institution of Structural Engineers (1955) Report on structural safety, *Structural Engineer*, 33: 141–149.

IWS (2002) *Proceedings of the International Workshop on "Foundation Design Codes and Soil Investigation in view of International Harmonization and Performance"*, Y. Honjo, O. Kusakabe, K. Matsui, M. Kouda and G. Pokharel (eds), Swets & Zeitlinger, Lisse.

Jaky, J. (1936) Stability of earth slopes, *Proceedings of the First International Conference on Soil Mechanics and Foundation Engineering*, Vol. 2, Cambridge, MA, pp. 125–129.

Jamiolkowski, M.B., Ricceri, G. and Simonini, P. (2009) Safeguarding Venice from high tides: site characterization and geotechnical problems, *Proceedings of the Seventeenth International Conference on Soil Management and Geotechnical Engineering*, Vol. 4, M. Hamza, M. Shahien and Y. El Mossallamy (eds), Millpress, Amsterdam, pp. 3209–3230.

Jardine, R.J. and Chow, F. (1996) *New Design Method for Offshore Piles*, MTD Pub. 96/103, Marine Technology Directorate, London.

Justason, M.D., Janes, M.C., Middendorp, P. and Mullins, A.G. (2000) Statnamic load testing using water as reaction mass, *Proceedings of the Sixth International Conference on the Application of Stress Wave Theory to Pile*, São Paulo.

Kant, I. (1793) *Über den gemeinspruch: Das mag in der Theorie richtig sein, taugt aber nicht für die Praxis, Kant's gesammelte Schriften*, Königliche Preussische Akademie der Wissenschaften, Berlin/Leipzig.

Katzenbach, R. and Turek, J. (2005) Combined pile-raft-foundation subjected to lateral loads, *Proceedings of the Sixteenth International Conference on Soil Management and Geotechnical Engineering, Vol. 4*, Osaka, pp. 2001–2004.

Kérisel, J.L. (1961) *Fondations profondes en milieux sableux. Proceedings of the Fifth International Conference on Soil Mechanics and Foundation Engineering, Vol. 2*, Paris, pp. 73–83.

Kim, J.B. and Sing, P.L. (1979) Pile cap soil interaction from full scale lateral load tests, *Journal of Geotechnical Engineering, ASCE*, 105(5): 643–653.

Kishida, H. (1967) Ultimate bearing capacity of piles driven into loose sand, *Soils and Foundations*, 7(3): 20–29.

Kitiyodom, P., Matsumoto, T. and Kanefusa, N. (2004) Influence of reaction piles on the behavior of a test pile in static load testing, *Canadian Geotechnical Journal*, 41: 408–420.

Koizumi, Y. and Ito, K. (1967) Field tests with regards to pile driving and bearing capacity of piled foundations, *Soils and Foundations*, 7(3): 30–53.

Kotthaus, M. and Jessberger, H.L. (1994) Centrifuge model tests on laterally loaded pile groups, *Proceedings of the Thirteenth International Conference on Soil Mechanics and Foundation Engineering, Vol. 2*, New Delhi, pp. 639–644.

Krishnan, R., Gazetas, G. and Velez, A. (1983) Static and dynamic lateral deflection of piles in non homogeneous soil stratum, *Géotechnique*, 33(3): 307–325.

Kulhawy, F.H. and Chen, Y.J. (1993) A thirty year perspective of Broms's lateral loading models as applied to drilled shafts, *B.B. Broms' Symposium on Geotechnical Engineering*, Singapore, pp. 225–240.

Kulhawy, F.H. and Goodman, R.E. (1980) Design of foundations on discontinuous rock, *Proceedings of the International Conference on Structural Foundations on Rock, Vol. 1*, Sydney, pp. 209–220.

Kulhawy, F.H. and Goodman, R.E. (1987) *Foundation in Rock*, Ground Engineering Reference Book, F.G. Bell (ed.), Butterworth, London, 55/1–55/13.

Kulhawy, F.H. and Phoon, K.K. (1993) *Drilled Shaft Side Resistance in Clay Soil to Rock*, Design and Performance of Deep Foundations, ASCE STP 38: 172–183.

Kulhawy, F.H., Beech, J.F. and Trautmann, C.H. (1989) Influence of geological development on horizontal stress in soils, *Foundation Engineering: Current Principles and Practices*, Evanston, IL.

Kuhlemeyer, R.L. (1979) Static and dynamic laterally loaded floating pile, *Journal of the Geotechnical Engineering Division, ASCE*, 105(2): 289–304.

Landi, G. (2006) *Piles under Horizontal Load: Experimental Investigations and Analysis*, PhD thesis, University of Naples [in Italian].

Landi, G. and Russo, G. (2005) *Interpretazione di prove sperimentali su pali sotto azioni orizzontali*, IARG, Ancona.

Lee, J. and Salgado, R. (1999) Determination of pile base resistance in sand, *Journal of Geotechnical and Geoenvironmental Engineering, ASCE*, 125(8): 673–683.

Lee, J., Salgado, R. and Paik, K. (2003) Estimation of the load capacity of pipe piles in sand based on CPT results, *Journal of Geotechnical and Geoenvironmental Engineering, ASCE*, 129(5): 391–403.

Leung, E.C. and Randolph, M.F. (2005) Finite element analysis of soil plug response, *International Journal for Numerical Analytical Methods in Geomechanics*, 15(2): 121–141.

Liu, J., Huang, Q., Li, H. and Hu, W.L. (1994) Experimental research on bearing behaviour of pile groups in soft soil, *Proceedings of the Thirteenth International Conference on Soil Mechanics and Foundation Engineering, Vol. 2*, New Delhi, pp. 535–538.

Liu, J., Yuan, Z.L. and Shang, P.K. (1985) Cap–pile–soil interaction of bored pile groups,

Proceedings of the Eleventh International Conference on Soil Mechanics and Foundation Engineering, Vol. 3, San Francisco, pp. 1433–1436.

Lopes, F.R. and Laprovitera, H. (1988) On the prediction of the bearing capacity of bored piles from dynamic penetration test, *Proceedings of the First International Conference on Foundations on Bored and Auger Piles*, BAP I, pp. 537–540.

Mandel, J. (1950) Tassements produits par la consolidation d'une couche d'argile de grande épaisseur, *Proceedings of the Fifth International Conference on Soil Mechanics and Foundation Engineering, Vol. 1*, Paris, pp. 733–756.

Mandolini, A. (1994) Modelling settlement behaviour of piled foundations, *Proceedings of a Workshop on Pile Foundations: Experimental Investigations, Analysis and Design*, Napoli, pp. 361–406.

Mandolini, A. and Viggiani, C. (1992) Terreni e opere di fondazione di un viadotto sul fiume Garigliano, *Rivista Italiana di Geotecnica*, 26(2): 95–114.

Mandolini, A. and Viggiani, C. (1997) Settlement of piled foundations, *Géotechnique*, 47(3): 791–816.

Mandolini, A., Price, G., Viggiani, C. and Wardle, I.F. (1992) Monitoring load sharing within a large pile cap foundation, *Proceedings of a Symposium in Paris "Géotechnique et Informatique"*, Paris, pp. 113–121.

Mandolini, A., Russo, G. and Viggiani, C. (1997) Pali per la riduzione dei cedimenti, *Atti delle Conferenze di Geotecnica di Torino XVI Ciclo*.

Mandolini, A., Russo, G. and Viggiani, C. (2005) Pile foundations: experimental investigations, analysis and design: state of the art report, *Proceedings of the Sixteenth International Conference on Soil Management and Geotechnical Engineering, Vol. 1*, Osaka.

McVay, M., Casper, R. and Shang, T.I. (1995) Lateral response of three-row groups in loose to dense sands at 3D and 5D pile spacing, *Journal of Geotechnical Engineering, ASCE*, 121(5): 436–441.

McVay, M., Zhang, L., Molnit, T. and Lai, P. (1998) Centrifuge testing of large laterally loaded pile groups in sand, *Journal of Geotechnical and Geoenvironmental Engineering, ASCE*, 124(10): 1016–1026.

Matlock, M.M. and Reese, L.C. (1956) Non-dimensional solutions for laterally loaded piles with soil modulus assumed proportional to depth, *Proceedings of the Eighth Texas Conference on Soil Mechanics and Foundation Engineering, Special Publication No. 29*, University of Texas, Austin, TX.

Matlock, H. and Reese, L.C. (1960) Generalized solutions for laterally loaded piles, *Journal of Soil Mechanics and Foundations Division, ASCE*, 86(5): 63–91.

Meyerhof, G.G. (1953) *Recherche sur la force portante des pieux*, Annales ITBTP, p. 371.

Meyerhof, G.G. (1959) Compaction of sand and bearing capacity of piles, *Journal of Soil Mechanics and Foundations Division, ASCE*, 85(6): 1–29.

Meyerhof, G.G. (1976) Bearing capacity and settlement of pile foundations, *Journal of Geotechnical Engineering, ASCE*, 102(3): 195–228.

Meyerhof, G.G. (1983) Scale effect of ultimate pile capacity, *Journal of Geotechnical Engineering, ASCE*, 109(6): 797–866.

Meyerhof, G.G. (1995) Behaviour of pile foundations under special loading conditions, *Canadian Geotechnical Journal*, 32(2): 204–222.

Meyerhof, G.G. and Valsangkar, A.J. (1977) Bearing capacity of piles in layered soils, *Proceedings of the Ninth International Conference on Soil Mechanics and Foundation Engineering, Vol. 1*, Tokyo, pp. 645–650.

Middendorp, P. (1993) First experiences with Statnamic load testing of foundation piles in Europe, *Proceedings of the Second International Geotechnical Seminar on Deep Foundations on Bored and Auger Piles*, BAP II, W.F. Van Impe (ed.), Balkema, Rotterdam, pp. 265–272.

Middendorp, P. and Bielefeld, M.W. (1995) Statnamic load testing and the influence of stress wave phenomena, *Proceedings of the First International Statnamic Seminar*, Vancouver, pp. 207–220.

Mindlin, R.D. (1936) Force at a point in the interior of a semi-infinite solid, *Physics*, 7(5): 195–202.

Mokwa, R.L. (1999) *Investigation of the Resistance of Pile Caps to Lateral Loading*, PhD thesis, Virginia Polytechnic Institute and State University, Blacksburg, VA.

Mori, G. (2003) Development of the screw steel pipe pile with toe wing "Tsubasa Pile", *Proceedings of the Fourth International Geotechnical Seminar on Deep Foundations on Bored and Auger Piles*, BAP IV, Millpress, Rotterdam.

Morrison, C. and Reese, L.C. (1986) *A Lateral Load Test of a Full-Scale Pile Group in Sand*, GR86–1, Federal Highway Administration, Washington DC.

Muskat, M. (1953) *The Flow of Homogeneous Fluids through Porous Media*, McGraw Hill, New York.

Mylonakis, G. and Gazetas, G. (1998) Settlement and additional internal forces of grouped piles in layered soils, *Géotechnique*, 48(1): 55–72.

Neely, W.J. (1991) Bearing capacity of augercast piles in sand, *Journal of Geotechnical Engineering, ASCE*, 117(2): 331–345.

Ochoa, M. and O'Neill, M.W. (1989) Lateral pile interaction factor in submerged sand, *Journal of Geotechnical Engineering, ASCE*, 115(3): 359–378.

O'Neill, M.W. and Reese, L.C. (1999) Drilled shafts: construction procedures and design methods, *Federal Highways Administration Report IF-99–025*, Washington, DC.

O'Neill, M.W., Hawkins, R.A. and Mahar, L.J. (1982) Load transfer mechanisms in piles and pile groups, *Journal of Geotechnical Engineering, ASCE*, 108(12): 1605–1623.

Ooi, P.S.K., Chang, B.K.F. and Wang, S. (2004) Simplified lateral load analysis of fixed head piles and pile groups, *Journal of Geotechnical and Geoenvironmental Engineering, ASCE*, 130(1): 1140–1151.

Osterberg, J.O. (1995) The Osterberg cell for load testing drilled shafts and driven piles, *Federal Highways Administration Research Report FHWA-SA-94-035*, pp. 1–92.

Osterberg, J.O. and Gill, S.A. (1973) Load transfer mechanism for piers socketed in hard soils or rocks, *Proceedings of the Ninth Canadian Symposium on Rock Mechanics*, Montreal, pp. 235–262.

Ovesen, N.K. (2002) Limit state design: the Danish experience, *Proceedings of the International Workshop on Foundation Design Codes and Soil Investigation*, pp. 107–116, Tokyo.

Pando, M., Filz, G., Hoppe, E., Ealy, C. and Muchard, M. (2000) Performance of a composite pile in a full scale Statnamic load testing program, *Proceedings of the Fifty-Third Canadian Geotechnical Conference, Vol. 1*, Montreal, pp. 909–916.

Peaker, K.R. (1984) Lakeview Tower: a case history of foundation failure, *Proceedings of the International Conference on Case Histories in Geotechnical Engineering*, S. Prakask (ed.), University of Missouri, Rolla, MO, pp. 7–13.

Pease, K.A. and Kulhawy, F.H. (1984) Load transfer mechanism in rock sockets and anchors, *Report EL-3777, Electric Power Research Institute*, Palo Alto, CA.

Peck, R.B. and Davisson, M.T. (1962) Discussion of "Design and stability considerations for unique pier" by J. Michalos and D.P Billington, *ASCE Transactions*, 127(IV): 413.

Polubarinova-Koschina, P. (1956) *Theory of Ground Water Movements*, Princeton University Press, Princeton, NJ.

Potts, D.M. and Zdravkovicz, L. (2001) *Finite Element Analysis in Geotechnical Engineering, Vol. 2: Applications*. Thomas Telford, London.

Poulos, H.G. (1968) Analysis of the settlement of pile groups, *Géotechnique*, 18(4): 449–471.

Poulos, H.G. (1971a) Behaviour of laterally loaded piles: I – single piles, *Journal of the Soil Mechanics and Foundations Division, ASCE*, 97(5): 711–731.

Poulos, H.G. (1971b) Behaviour of laterally loaded piles: I – pile groups, *Journal of the Soil Mechanics and Foundations Division, ASCE*, 97(5): 733–751.

Poulos, H.G. (1971c) Behaviour of laterally loaded piles: II – pile groups, *Journal of the Soil Mechanics and Foundations Division, ASCE*, 97(5): 733–751.

Poulos, H.G. (1972a) Behavior of laterally loaded piles: III – socketed piles, *Journal of the Soil Mechanics and Foundations Division, ASCE*, 98(4): 341–360.

Poulos, H.G. (1972b) Load–settlement prediction for piles and piers, *Journal of the Soil Mechanics and Foundations Division, ASCE*, 98(9): 879–897.

Poulos, H.G. (1973) Analysis of piles in soil undergoing lateral movement, *Journal of the Soil Mechanics and Foundations Division, ASCE*, 99(5): 391–406.

Poulos, H.G. (1988) Modified calculation of pile group settlement interaction, *Journal of Geotechnical Engineering, ASCE*, 114(6): 697–706.

Poulos, H.G. (1994) An approximate numerical analysis of pile raft interaction, *International Journal of Analytical and Numerical Methods in Geomechanics*, 18(2): 73–92.

Poulos, H.G. (2000) Pile testing from the designer's viewpoint, *Statnamic Loading Test '98*, Balkema, Rotterdam, pp. 45–55.

Poulos, H.G. (2008) Simulation of the performance and remediation of imperfect pile groups, *Proceedings of the Fifth International Symposium on Deep Foundations on Cored and Auger Piles*, W.F. Van Impe and P.O. Van Impe (eds), CRC Press, Boca Raton, FL.

Poulos, H.G. and Davis, E.H. (1968) The settlement behaviour of single axially loaded incompressible piles and piers, *Géotechnique*, 18(3): 351–371.

Poulos, H.G. and Davis, E.H. (1980) *Pile Foundations Analysis and Design*, John Wiley and Sons, New York.

Poulos, H.G., Carter, J.P. and Small J.C. (2001) Foundations and retaining structures: research and practice, *Proceedings of the Fifteenth International Conference on Soil Mechanics and Geotechnical Engineering, Vol 4*, Istanbul, pp. 2527–2606.

Prandtl, L. (1921) *Über die Harte plastischer Körper*, Nachrichten von der Gesellschaft der Wissenschaften zu Goettingen, Matemathik-Physikalische Klasse, 37: 74–85.

Price, G. and Wardle, I.F. (1982) A comparison between cone penetration test results and the performance of small diameter instrumented piles in stiff clays, *Proceedings of the Second European Symposium on Penetration Testing Vol. 2*, Amsterdam, pp. 775–780.

Radhakrishnan, R. and Leung, C.F. (1989) Load transfer behavior of rock-socketed piles, *Journal of Geotechnical Engineering, ASCE*, 115(6): 755–768.

Randolph, M.F. (1981) Response of flexible piles to lateral loading, *Géotechnique*, 31(2): 247–259.

Randolph, M.F. (1983) Design considerations for offshore piles, *Proceedings of the ASCE Special Conference on Geotechnical Practice in Offshore Engineering*, Austin, TX, pp. 422–439.

Randolph, M.F. (1987) *PIGLET: A Computer Program for the Analysis and Design of Pile Groups*, Report GEO 87036, Perth, University of Western Australia.

Randolph, M.F. (1989) *RATZ: Load Transfer Analysis of Axially Loaded Piles*, Research Report No. 86003, Perth, University of Western Australia, Perth.

Randolph, M.F. (1994) Design methods for pile groups and piled rafts, state-of-the-art report, *Proceedings of the Thirteenth International Conference on Soil Mechanics and Foundation Engineering, Vol. 5*, New Delhi, pp. 61–82.

Randolph, M.F. (1998) Modelling of offshore foundations, part 2: anchoring systems, E.H. Davis Memorial Lecture, *Australian Geomechanics*, 33(3): 3–16.

Randolph, M.F. (2003) Science and empiricism in pile foundations design, *Géotechnique*, 53(10): 847–875.

Randolph, M.F. and Clancy, P. (1993) Efficient design of piled rafts, *Proceedings of the Second International Geotechnical Seminar on Deep Foundations on Bored and Auger Piles*, BAP II, W.F. Van Impe (ed.), Balkema, Rotterdam, pp. 119–130.

Randolph, M.F. and Murphy, B.S. (1985) Shaft capacity of driven piles in clay, OTC Report 4883, *Proceedings of the Seventeenth Annual Offshore Technology Conference, Vol. 1*, Houston, TX, pp. 371–378.

Randolph, M.F. and Poulos, H.G. (1982) Estimating the flexibility of offshore pile groups, *Proceedings of the Second International Conference on Numerical Methods in Offshore Piling*, University of Texas, Austin, pp. 313–328.

Randolph, M.F. and Wroth, C.P. (1978) Analysis of deformation of vertically loaded piles, *Journal of the Geotetchnical Engineering Division, ASCE*, 104(12): 1465–1488.

Randolph, M.F. and Wroth, C.P. (1979) An analysis of the vertical deformation of pile groups, *Géotechnique*, 29(4): 423–439.

Recinto, B. (2004) *Sperimentazione in vera grandezza sui pali di fondazione. Modalità di prova e interpretazione*, PhD thesis, Università di Napoli Federico II.

Reese, L.C. and O'Neill, M.W. (1988) Drilled shafts: construction and design, *Federal Highways Administration Report No. HI-88-042*, Washington, DC.

Reese, L.C. and O'Neill, M.W. (1989) New design method for drilled shafts from common soil and rock tests, *Foundation Engineering: Current Principles and Practices*, Evanston, IL, II: 1026–1039.

Reese, L.C. and Van Impe, W.F. (2001) *Single Piles and Pile Groups under Lateral Loading*, Balkema, Rotterdam.

Reese, L.C., Cox, W.R. and Koop, F.D. (1974) Analysis of laterally loaded piles in sand, OTC Report 2080, *Proceedings of the Sixth Offshore Technology Conference*, Houston, TX, pp. 473–483.

Reese, L.C., Cox, W.R. and Koop, F.D. (1975) Field testing and analysis of laterally loaded piles in stiff clay, *Proceedings of the Seventh Annual Offshore Technology Conference*, Vol. 2, Houston, TX, pp. 672–690.

Reese, L.C., Isenhower, W.M. and Wang, S.T. (2006) *Analysis and Design of Shallow and Deep Foundations*, John Wiley and Sons, Hoboken, NJ.

Reul, O. (2004) Numerical study of the bearing behavior of piled rafts, *International Journal of Geomechanics, ASCE*, 4(2): pp. 59–68.

Rollins, K.M. and Sparks, A. (2002) Lateral resistance of full-scale pile cap with gravel backfill, *Journal of Geotechnical and Geoenvironmental Engineering, ASCE*, 128(9): 711–723.

Rollins, K.M., Lane, J.D. and Gerber, T.M. (2005) Measured and computed lateral response of a pile group in sand, *Journal of Geotechnical and Geoenvironmental Engineering, ASCE*, 131(1): 103–114.

Rollins, K.M., Peterson, K.T. and Weaver, T.J. (1998) Lateral load behavior of full-scale pile group in clay, *Journal of Geotechnical and Geoenvironmental Engineering, ASCE*, 124(6): 468–478.

Rowe, R.K. and Armitage, H.H. (1987) Theoretical solutions for axial deformation of drilled shafts in rock, *Canadian Geotechnical Journal*, 24(1): 114–125.

Ruesta, P.F. and Townsend, F.C. (1997) Evaluation of laterally loaded pile group at Roosevelt Bridge, *Journal of Geotechnical and Geoenvironmental Engineering, ASCE*, 123(12): 1153–1161.

Russo, G. (1994) Monitoring the behaviour of a pile foundation, *Proceedings of a Workshop on Pile Foundations: Experimental Investigations, Analysis and Design*, Napoli, pp 435–441

Russo, G. (1996) Soil structure interaction for piled rafts, PhD thesis, University of Napoli Federico II [in Italian].

Russo, G. (1998a) Numerical analysis of piled rafts, *International Journal of Analytical and Numerical Methods in Geomechanics*, 22(6): 477–493.

Russo, G. (1998b) Developments in the analysis and design of piled rafts, *Proceedings of a Workshop on Prediction and Performance in Geotechnical Engineering*, Hevelius, Benevento.

Russo, G. and Viggiani, C. (1995) Long term monitoring of a pile foundation, *Proceedings of the Fourth International Symposium of Field Measurements in Geomechanics*, Bergamo.

Russo, G. and Viggiani, C. (1997) Some aspects of numerical analysis of piled rafts, *Proceedings of the Fourteenth International Conference on Soil Mechanics and Foundation Engineering*, Hamburg.

Russo, G. and Viggiani, C. (2008) Piles under horizontal load: an overview, *Proceedings of the Second International British Geotechnical Association Conference on Foundations*, ICOF 2008, ed. M.J. Brown, M.F. Bransby, A.J. Brennan and J.A. Knappett, BREPress, Watford.

Salgado, R. (2006) Analysis of the axial response of non-displacement piles in sand, Geomechanics II: Testing, modelling and simulation, *Proceedings of the Second US-Japan Workshop, ASCE Special Geotechnical Publication No. 143*, pp. 427–439.

Salgado, R. (2008) *The Engineering of Foundations*, McGraw-Hill, New York.

Schmertmann, J.H. (1978) Guidelines for cone penetration test, performance and design, *Federal Highways Adminstration Report No. FHWA-TS-78-209*, Washington DC.

Schmertmann, J.H. and Hayes J.A. (1997) The Osterberg cell and bored pile testing: a symbiosis, *Proceedings of the Third International Geotechnical Engineering Conference*, Cairo University, pp. 139–166.

Schmertmann, J.H., Hartman, J.D. and Brown, P.R. (1978) Improved strain influence factor diagrams, *Journal of the Geotechnical Engineering Division, ASCE*, 104(8): 1131–1135.

Schmidt, H.G. (1981) Group action of laterally loaded bored piles, *Proceedings of the Tenth International Conference on Soil Mechanics and Foundation Engineering, Vol. 2*, Stockholm, pp. 833–837.

Scott, R.F. (1981) *Foundation Analysis*, Civil Engineering and Engineering Mechanics Series, Prentice-Hall, Englewood Cliffs, NJ.

Seidel, J.P. and Haberfield, C.M. (1995) The axial capacity of pile sockets in rocks and hard soils, *Ground Engineering*, 28(2): 33–38.

Seidel, J.P. and Cho, C.W. (2002) Prediction of the shaft resistance of pile sockets, *Journal of the Korean Geotechnical Society*, 8(5): 281–293.

Selby, A.G. and Poulos, H.G. (1984) Lateral load tests on model pile groups, *Proceedings of the Fourth Australia-New Zealand Conference on Geomechanics*, Perth, pp. 154–158.

Skempton, A.W. (1954) The pore pressure coefficients A and B, *Géotechnique*, 4(4): 143–147.

Skempton, A.W. and Bjerrum, L. (1957) A contribution to the settlement analysis of foundations on clay, *Géotechnique*, 7(4): 168–178.

Skempton, A.W., Yassin, A.A., Gibson, R.E. (1953) Théorie de la force portante des pieux dans la sable, *Annales ITBTP*, pp. 285–290.

Smith, E.A.L. (1960) Pile-driving analysis by the wave equation, *Journal of Soil Mechanics and Foundation Engineering, ASCE*, 86(4): 63–75.

Sørensen, T. and Hansen, B. (1956) Pile driving formulae: an investigation based on dimensional considerations and statistical analysis, *Proceedings of the Fourth International Conference on Soil Mechanics and Foundation Engineering, Vol. 2*, London, pp. 61–65.

Sowers, G.F. (1979) *Introductory Soil Mechanics and Foundations: Geotechnical Engineering*, fourth edition, Macmillan, New York.

Spillers, W.R. and Stoll, R.D. (1964) Lateral response of piles, *Journal of the Soil Mechanics and Foundations Division, ASCE*, 90(6): 1–9.

Terzaghi, K. (1924) *Die Berechnung der Durchlässigkeitziffer des Tones aus dem Verlauf der hydrodinamischen Spannungserscheinungen*, Sitzungberichte der Akademie der Wissenschaften in Wien, matematisch-naturwissenschfterliche, Klasse, Abteilung IIa, 132: 125–138, Vienna.

Terzaghi, K.(1943) *Theoretical Soil Mechanics*, John Wiley, New York.

Terzaghi, K. and Peck, R.B. (1948) *Soil Mechanics in Engineering Practice*, John Wiley, New York.

Thorburn, S. and MacVicar, R.S.L. (1971) Pile load tests to failure in the Clyde alluvium, *Proceedings of the Conference on the Behaviour of Piles*, ICE, London, pp. 1–7.

Thornburn, S., Laird, C. and Randolph, M.F. (1983) Storage tanks founded on soft soils reinforced with driven piles, *Proceedings of the Conference on Recent Advances in Piling and Ground Treatment*, ICE, London, pp. 157–164.

Tomlinson, M.J. (1987) *Pile Design and Construction Practice*, third edition, Viewpoint Publications, London.

Tomlinson, M.J. (1994) *Pile Design and Construction Practice*, fourth edition, E. & F.N. Spon, London.

Van Impe, W.F. (1991) Deformation of deep foundations, *Proceedings of the Tenth European Conference on Soil Mechanics and Foundation Engineering*, Vol. 3, Florence, pp. 1031–1064.

Van Impe, W.F. (2003) Belgian geotechnics' experts research on screw piles, *Belgian Screw Pile Technology: Design and Recent Developments*, ed. J. Maertens and N. Huybrechts, Balkema, Rotterdam, pp. xiii–xvii.

Van Impe, W.F. and Lungu, I. (1996) *Technical Report on Settlement Prediction Methods for Piled Raft Foundations*, Ghent University, Belgium.

Van Impe, W.F., Viggiani, C., Van Impe, P.O., Russo, G. and Bottiau, M. (1998) Load–settlement behaviour versus distinctive Omega-pile execution parameters, *Proceedings of the Third International Conference on Foundations on Bored and Auger Piles, BAP III*, W.F. Van Impe (ed.), Balkema, Rotterdam, pp. 355–368.

Vaughan, P.R., Kovacevic, N. and Potts, D.M. (2004) Then and now: some comments on the design and analysis of slopes and embankments, *The ICE Skempton Conference on Advances in Geotechnical Engineering, Vol. 3*, Thomas Telford, London, pp. 15–64.

Vesic, A.S. (1964) Investigation of bearing capacity of piles in sand, *Proceedings of the North American Conference on Deep Foundations, Vol. 1*, Mexico City, pp. 197–224.

Vesic, A.S. (1968) Experiments with instrumented pile groups in sands, *Performance of Deep Foundations*, ASTM Special Technical Publication No. 444, p. 177–222.

Vesic, A.S. (1977) *Design of Pile Foundations: National Cooperative Highway Research Program, Synthesis Highway Practice Report No. 42*, Transport Research Board, Washington, DC.

Viggiani, C. (1993) Further experiences with auger piles in Naples area, *Proceedings of the Second International Geotechnical Seminar on Deep Foundations on Bored and Auger Piles, BAP II*, W.F. Van Impe (ed.), Balkema, Rotterdam, pp. 445–458.

Viggiani, C. (1998) Pile groups and piled rafts behaviour, *Proceedings of the Third International Geotechnical Seminar on Deep Foundations on Bored and Auger Piles, BAP III*, W.F. Van Impe and W. Haegman (eds), Balkema, Rotterdam, pp. 77–94.

Viggiani, C. (1999) *Fondazioni*, Hevelius, Benevento.

Viggiani, C. (2001) Analysis and design of piled foundations, First Arrigo Croce Lecture, *Rivista Italiana di Geotecnica*, 35(1): 47–75.

Viggiani, C. and Viggiani, G.M.B. (2008) Il ruolo dell'osservazione delle opere nell'Ingegneria Geotecnica, *Diagnostica per la tutela e la conservazione dei materiali nel costruito*, Diacomast 2, Acos srl, S. Leucio.

Wardle, L.J. and Fraser, R.A. (1974) Finite element analysis of a plate on a layered cross-anisotropic foundation, *Proceedings of the First International Conference of Finite Element Methods in Engineering*, University of New South Wales, Australia, pp. 565–578.

Wellington, A.M. (ed.) (1893) *Piles and Pile-Driving*, Engineering News Publishing Co., New York.

White, D.J. (2003) PSD measurement using the single particle optical sizing (SPOS) method, *Géotechnique*, 53(3): 317–326.

Williams, A.F. and Pells, P.J.N. (1981) Side resistance of rock sockets in sandstone, mudstone and shale, *Canadian Geotechnical Journal*, 18: 502–513.

Wilson, S.D. and Hilts, D.E. (1967) How to determine lateral load capacity of piles, *Wood Preserving News*, July.

Wong, I.H., Chang, M.F. and Cao, X.D. (2000) Raft foundations with disconnected settlement-reducing piles, J.A. Hemsley (ed.), *Design Application of Raft Foundations and Ground Slabs*, Thomas Telford, London, Chapter 17, pp. 469–486.

Yang, Z. and Jeremić, B. (2002) Numerical analysis of pile behaviour under lateral loads in layered elastic-plastic soils, *International Journal for Numerical and Analytical Methods in Geomechanics*, 26(14): 1385–1406.

Yang, Z. and Jeremić, B. (2003) Numerical study of group effects for pile groups in sands, *International Journal for Numerical and Analytical Methods in Geomechanics*, 27(15): 1255–1276.

Yasufuku, N., Ochiai, H. and Maeda, Y. (1997) Geotechnical analysis of skin friction of cast-in-place piles, *Proceedings of the Fourteenth International Conference on Soil Mechanics and Foundation Engineering, Vol. 2*, Hamburg, pp. 921–924.

Zafir, Z. and Vanderpool, W.E. (1998) Lateral response of large diameter drilled shafts: I-15/US 95 load test program, *Proceedings of the Thirty-Third Engineering Geology and Geotechnical Engineering Symposium*, University of Nevada, Reno, pp. 161–176.

Zhang, L. and Einstein, H.H. (1998) End bearing resistance of drilled shafts in rock, *Journal of Geotechnical and Geoenvironmental Engineering*, 124(7): 574–584.

Zienkiewicz, O.C. and Taylor, R.L. (1991) *The Finite Element Method Vol. 2: Solid and Fluid Mechanics, Dynamics and Non-linearity*, fourth edition, McGraw-Hill, London.

Index

Printed and bound by CPI Group (UK) Ltd, Croydon, CR0 4YY

01/11/2024

01782605-0004